IN THE COMPANY OF CARS

Human Factors in Road and Rail Transport

Series Editors

Dr Lisa Dorn
Director of the Driving Research Group, Department of Human Factors,
Cranfield University

Dr Gerald Matthews
Professor of Psychology at the University of Cincinnati

Dr Ian Glendon
Associate Professor of Psychology at Griffith University, Queensland,
and is president of the Division of Traffic and Transportation Psychology
of the International Association of Applied Psychology

Today's society must confront major land transport problems. The human and financial costs of vehicle accidents are increasing, with road traffic accidents predicted to become the third largest cause of death and injury across the world by 2020. Several social trends pose threats to safety, including increasing car ownership and traffic congestion, the increased complexity of the human-vehicle interface, the ageing of populations in the developed world, and a possible influx of young vehicle operators in the developing world.

Ashgate's 'Human Factors in Road and Rail Transport' series aims to make a timely contribution to these issues by focusing on the driver as a contributing causal agent in road and rail accidents. The series seeks to reflect the increasing demand for safe, efficient and economical land-based transport by reporting on the state-of-theart science that may be applied to reduce vehicle collisions, improve the usability of vehicles and enhance the operator's wellbeing and satisfaction. It will do so by disseminating new theoretical and empirical research from specialists in the behavioural and allied disciplines, including traffic psychology, human factors and ergonomics.

The series captures topics such as driver behaviour, driver training, in-vehicle technology, driver health and driver assessment. Specially commissioned works from internationally recognised experts in the field will provide authoritative accounts of the leading approaches to this significant real-world problem.

In the Company of Cars
Driving as a Social and Cultural Practice

SARAH REDSHAW

*Honorary Associate, Department of Psychology, Macquarie University,
Sydney, Australia*

CRC Press
Taylor & Francis Group
Boca Raton London New York

CRC Press is an imprint of the
Taylor & Francis Group, an **informa** business

CRC Press
Taylor & Francis Group
6000 Broken Sound Parkway NW, Suite 300
Boca Raton, FL 33487-2742

First issued in paperback 2017

© 2008 by Sarah Redshaw
CRC Press is an imprint of Taylor & Francis Group, an Informa business

No claim to original U.S. Government works

Version Date: 20160226

ISBN 13: 978-0-7546-7198-5 (hbk)
ISBN 13: 978-1-138-07179-7 (pbk)

Visit the Taylor & Francis Web site at
http://www.taylorandfrancis.com

and the CRC Press Web site at
http://www.crcpress.com

Contents

List of Figures

Foreword

Driving may have long since ceased to be a glamorous or special activity and driving a petrol or diesel engined car seems to have become somewhat antisocial since concern about their CO_2 emissions and the effect on climate change has become an issue. But there has been no sign – yet – that people are going to stop driving cars. In 2005 in the United Kingdom 678 billion passenger kilometres were done in cars, vans and taxis. This compares with only 52 billion passenger kilometres by rail and 10 billion by air (ONS 2007, 158). Travel by bicycle, by coach and by bus may be in decline but there continues to be an increase in car travel, the mode of transport that has contributed most to the massive increase in personal transport since 1961. This pattern of increase in mobility led by car travel is characteristic of the western industrialised world; people want to drive!

Of course travelling in cars does not necessarily involve driving – some people are simply passengers in other people's cars, limousines or taxis – and what driving itself involves is rapidly changing. A major European manufacturer recently sent out some advertising puff for a new model that is, they suggest, 'the car that sees things before you do'… Cars have been taking over the physical aspects of driving for some time. Power assisted brakes are just about ubiquitous now and power assisted steering very nearly so. Automatic gearboxes are smooth and work well even on lower powered cars. Sensors in the car have long channelled information to instruments on the dashboard about speed, engine 'revs', oil pressure, the amount of fuel remaining and the functioning of electrical equipment. Now sensors on some cars are linked to electronic systems that warn of brakes wearing, falling tyre pressures and can calculate how many miles worth of fuel is left as well as when the next service is due. On-board sensor systems can give audible advice on whether the seat belt has been fastened or how close the car is getting to whatever is behind when parking. Car navigation systems have taken over the hassle of map reading … for the most part.

These automatic systems have begun to take over first the manual functions of driving and then progressively the programmatic aspects of what André Leroi-Gourhan (1993, 230) called the 'chaîne operatoire', the operational sequence through which human gestures are organised in relation to tools and objects to bring about effects in the material world. He pointed out that unlike other animal species, human evolution happens outside the body. For most species, tool and gesture evolved together within the body; the protective shell, the grasping claw, the accelerating hind legs. But in human evolution the tool has become externalised and separated from the gesture which operates it. As he puts it 'tools were 'exuded' by humans in the course of their evolution' as fists and nails were replaced by hammers and knives (Leroi-Gourhan 1993, 239). The motor car is itself such a tool, evolved to replace the harnessing of horses to carriages, but within the car there are a set of tools that are used to achieve 'driving' of the car; the steering wheel, brakes, accelerator and other controls. In Leroi-Gourhan's account of human technological evolution, 'motor' functions are transferred to the machine (e.g. power assisted brakes and steering). Then the sequence of actions in the machine becomes progressively automated (e.g. automatic gears). This sort of automation requires programme memory that can be mechanical (the odometer, the automatic gearbox) but can also be electronic (the on-board computer that tracks the workings of subsystems). Finally, the machine takes control of the whole sequence as the programme that ultimately determines the operational sequence of driving itself is recorded and managed within the tool (the self-driving car). This begins with the car knowing when to change gear and then extends to include turning on the lights or the windscreen wipers in response to the environment. Many cars know how to control the braking and the torque to avoid a skid and they take over these driving duties without the driver even noticing. These automatic functions leave the driver free to think, to listen to the radio or music system, to make phone calls (hands free of course). Just how long will it be before the human driver is made completely redundant?

Leroi-Gourhan argues that the major part of individual behaviour has become 'mechanical' as it has been acquired through the culture and has become routine. This is of course true of driving. What was once an opportunity to experience the sensation of enhanced motor control that could bring pleasure through the skillful manipulation of the direction and speed of the car has become, for the most part, a routine, mechanical activity of driving within the confines of white lines, speed limits and the constraints of traffic. As long ago as 1947 Max Horkheimer said of the car that the freedom it brought also brought a change in what freedom was: 'There are speed limits, warnings to drive slowly, to stop, to stay within certain lanes, and even diagrams showing the shape of the curve ahead. We must keep our eyes on the road and be ready at each instant to react with the right motion. Our spontaneity has been replaced by a frame of mind which compels us to discard every emotion or idea that might impair our alertness to the impersonal demands assailing us' (1947, 98). Put like that, no wonder some of us are keen to get the machine to take over driving and let us do something more interesting. In 2005 a Volkswagen Toureg known as 'Stanley' drove itself to victory in the 'Grand Challenge' across the Mojave Desert in the United States organised by the Defence Advanced Research Projects Agency (DARPA). In 2007 it was a Chevy Tahoe called 'Boss' that won the urban version of the challenge which involved the driverless vehicles not just getting where they were going as fast as possible, but also having to obey all traffic rules and avoid other vehicles. In 2004 no car could complete the original desert 'Grand Challenge' – driverless cars have learnt very fast to undertake what we humans have long thought is a distinctively demanding mix of motor, sensory and moral skills. Driving in contemporary societies leaves little room for spontaneity – for many, if not most, of us, the arrival of cars that can safely but conveniently and efficiently move us around within the constraints of traffic and the road system will be welcomed. Unlike public transport, the driverless car should at least leave us with the freedom of deciding when and where to go.

However, driverless cars for the mass consumer are still at least twenty to thirty years off and while cars will be progressively 'autonomised' over that period, the pleasures and tribulations of driving and its social costs and benefits will continue. What is remarkable is that within such an ordered and managed road system as one finds in industrialised countries today, there is considerable variation in ways of driving. The desire to drive faster than the limits (at least between speed cameras…), or insist on driving slowly whatever the general pace is, the willingness or not to undertake manoeuvres such as squeezing through a narrow gap or parking in a tight space, the way the car is placed on the road relatively to other cars, the ease with which road space is given up or grabbed – all of these things remain variable between drivers and a matter of concern to all road users. Of course most of us have a moral certainty about *our own* actions that gives us the right to identify and criticise those of others we identify as deviants; incompetent, aggressive, mean, careless, selfish, intolerant. It is the cultural and social dimensions of human interaction with the car that is the focus of Sarah Redshaw's book which examines driving from the aggressive practices of speed to the ethics of pedestrian crossings and roundabouts, from the gender differences in driving behaviours to the strategies of manufacturers in making and selling cars to drivers. This comprehensive overview of what is entailed in living 'in the company of cars' draws together a range of sources and materials to think through the evolving dynamic relationship between driver and car. It will serve as a key resource for those curious about what is entailed in driving … at least until driving becomes no more than a hobby and a pastime once again, just as horse riding did for those living in the urbanised and industrial world during the early part of the twentieth century.

Tim Dant

References

Horkheimer, Max (1974 [1947]) *The Eclipse of Reason*, New York: Continuum Books.

Leroi-Gourhan, André (1993 [1964]) *Gesture and Speech*, Cambridge, Massachusetts: MIT Press.

ONS Office for National Statistics (2007) *Social Trends 37*, Houndmills, Basingstoke: Palgrave Macmillan.

Preface

In the Company of Cars is intended to give a social/cultural perspective on driving in order to provide some qualitative insight as well as background context to the plethora of psychological and epidemiological research in the area of road safety. It is also intended as a detailed investigation of issues raised in the relatively new area of mobilities research which stems largely from sociology. Mobilities research does not simply provide a social science perspective in an area dominated by the hard sciences, it goes much further in considering the deeper social and cultural forms and relationships influencing and informing mobility. Important theorisations in this area have provided novel and exciting ways of looking at automobility as well as other kinds of mobility that can be informative in planning and transportation policy.

There is a slippage between the social and cultural in the book but allowing this slippage was considered preferable to attempting to clearly delineate territories that do overlap in many ways. Nevertheless the social is not reducible to the cultural nor the cultural to the social. Each denotes important facets of human engagement that are not subsumed in the other.

The main aim of the book has been to give some detailed consideration to important social and cultural dimensions of driving in a way that has not been done before, drawing on concrete experiences wherever possible. The importance of social and cultural factors does not diminish the agency of individuals in their engagement with cars but it does indicate the extent to which the parameters of individual engagement are delineated by broader factors. The agency of individuals is nevertheless apparent in the investigation of experiences of driving and every attempt has been made to draw on theorisations that favour the acknowledgement of agency. Striking a balance between individual agency and social and cultural forces is difficult and no doubt not perfected here since it is the ongoing task of social theory and there is much more to learn. The balance between theory and concrete experience is also important and one that I have attempted to accomplish in the book though again this is an ongoing task requiring contributions from many different directions.

Acknowledgements

The majority of the research on which this book draws was conducted while I was a Postdoctoral Research Fellow at the Centre for Cultural Research, University of Western Sydney 2003–2006. The main project *Transforming Drivers: Driving as a Social, Cultural and Gendered Practice*, for which I was principal researcher was a partnership between the university and motoring organisation NRMA Motoring and Services. The team of investigators consisted of Zöe Sofoulis and Greg Noble from the University of Western Sydney and partner investigators Alan Finlay and Anne Morphett. The research was reported in Sofoulis, Zöe, Noble, Greg and Redshaw, Sarah (2005) *Youth Media and Driving Messages*, Transforming Drivers Report 1A. Sydney: NRMA Motoring and Services and Redshaw, Sarah and Noble, Greg (2006) *Mobility, Gender and Young Drivers: Second Report of the Transforming Drivers Study of Young People and Driving*, Sydney: NRMA Motoring and Services (available at http://www.uws.edu.au/research/researchcentres/ccr/publications#3).

As Project Director of the Driving Cultures Research I was engaged in a number of projects that have contributed to the research culminating in this book including projects with local and state government departments. Innovative city councils with a commitment to young driver initiatives included Blue Mountains, Penrith, Hawkesbury, Fairfield and Campbelltown Councils. The initial Driving Cultures project was funded with an innovative projects grant from the New South Wales Attorney General's Department and followed up with a project piloting educational programs for young people in schools with the Youth and School Programs Section of the New South Wales Roads and Traffic Authority. Exchanges with all the people involved in these projects and the young people who participated have all been invaluable for the research.

Thanks to Professor Ien Ang, Director, Centre for Cultural Research for the duration of the Driving Cultures program of research and the Director of Research Services at the university Gar Jones for their support of initial and subsequent grant proposals. The intellectual exchange possible with other researchers including Fiona Allon, Sharon Chalmers, Greg Gow, Anne Hurni and Elizabeth Cassity through the Centre for Cultural Research was stimulating and challenging. Professor Graeme Turner, Director of the Centre for Critical and Cultural Research at the University of Queensland has offered invaluable advice and support. Exchanges and collaborations with Fiona Nicoll now at the University of Queensland have been and continue to be significant and important to the continuance of this work providing many invaluable opportunities to talk through ideas.

A visit to the Centre for Mobilities Research at Lancaster University in Britain was a highlight of my Postdoctoral Fellowship and I was fortunate enough to attend the Mobilising Hospitality workshop in September 2005. I have drawn on the papers given at the workshop in my final chapter and must particularly acknowledge Professor Ghassan Hage for his stimulating keynote address at the workshop. Subsequently meeting Professor John Urry from the Centre was another valuable encounter. Professor Urry's work in particular has helped to situate this research within the broad area of mobility rather than road safety, giving it an emphasis that is more connected to social and cultural research.

I am grateful also for the opportunities for publication that have contributed to the development of the thoughts expressed in this book. Thanks to the publishers, editors and referees involved. These include:

2006 Acceleration: The Limits of Speed in *The Reinvention of Everyday Life: Culture in the Twenty-first Century*, edited by Howard McNaughton and Adam Lam. University of Canterbury Press, Christchurch, 195–206.

2007 Articulations of the Car: the dominant articulations of racing and rally driving, *Mobilities*, 2,1, 121–141.

2006 Driving Cultures: Cars, Young People and Cultural Research, *Cultural Studies Review*, 12:2, 74–89.

2005 Developing Control amongst Young Drivers, *Youth Studies Australia*, 24:3, 37–41.

2004 Young People's Ideas on Speed. *Journal of Road and Transport Research*, 13:4, 51–62.

2001 Changing Driving Behaviour – a cultural approach, *Australian Journal of Social Issues*, 36:4, 315–331.

The writing of the book has been supported by the Department of Psychology at Macquarie University through my being appointed an Honorary Associate for 2007/2008 allowing me access to the library and online journals, an invaluable resource for an undertaking such as this. Thanks especially to Dr Julia Irwin and Associate Professor Julie Fitness from the department.

Being part of an organisation such as the Australasian College of Road Safety has included me in conversations critical to the research. Thanks to Professor Mark Stevenson of the George Institute at Sydney University, Secretary Ian Faulks and in his role with the New South Wales Parliamentary Staysafe Committee and current President Liz de Rome for this opportunity. Other important mentors have included Niki Harre in the Department of Psychology at Auckland University, Malcolm Vick at James Cook University and Michael Fine, Department of Sociology at Macquarie University.

Many friends have supported me through the process of the research and the writing of the book, in particular Kath Harrison who has been my stable support throughout, Puck and Leta, Bramani, Fiona and Sandi, my sister Ruth and Libby who has read the manuscript with great interest and other family members who have seen the project develop. Thanks to Debbie Beaumont for ongoing mentoring and support. My nephews Anthony and Mark have supplied many examples of young men and their handling of cars as I have seen them grow up and mature. Thankfully their exploits left them unharmed.

My father Allan Redshaw was involved in the computerisation of the transport industry in Australia in the late sixties. Working for TNT he made several visits to the United States to look at the management of the transport industry there. He died in 1970 when I was 12 years old in a car crash, driving between Adelaide and Sydney. He was driving too fast and went to sleep at the wheel in a powerful car and cars were less forgiving in those days. Losing my father at such a young age has made the consequences and costs of road crashes very real and personal for me.

Some of the photographs in the book were taken by my uncle Ken Redshaw who died in 1985. Ken was an aerial roads photographer in the 1960s and I am very grateful for the opportunity to use some of his work in the book. Thanks to his wife Marlies for supplying me with the negatives and for her enthusiasm in having the photographs included here. Many thanks to Max Hill for his able assistance with formatting image files.

Dedicated to my father Allan Redshaw (1931–1970) and my uncle Ken Redshaw (1929–1985).

Introduction

Cars and their Associations

Western consumer cultures are very attached to private cars and particular cars are seen as saying something about those who drive them. They are also driven in ways that express the particularity of both the car and the driver. Car cultures intersect with driving cultures within car networks that include how cars are embedded in socialisation. While cars have been regarded as neutral technology the driver has been blamed for mishaps and disasters. The book raises the importance of considering the interaction between car and driver suggesting that culpability rests with this interaction, and not solely with the driver. In this context, cars are considered as dangerous machines that are articulated in particular ways related to the type of car as well as characteristics of the driver.

Driving then, is a cultural practice that becomes routinised, that is, it becomes regulated or habitual according to accepted and agreed ways of behaving that are characteristic of the society and particular cultures within it. Driving as a practice is not purely individualised but limited to and by shared cultural meanings, associations and practices. Driving is above all a social practice even though it is highly privatised with individuals separated from each other in their separate cars. Each nevertheless relies on a public system of roads and socially agreed methods of operating that include the road rules and other stipulations of the authorities that govern them. Driving is routinised in being regulated by social and cultural values, forms of embodiment and behaviour and its connection to other aspects of culture such as gender. Driving is not, however, a uniform practice and there are many shared meanings connected with different cultural groups. There are thus driving cultures, as there are car cultures, with many variations and dissimilarities from each other in the values, beliefs, expectations and embodiments related to cars and driving.

Three major aspects of driving through which it becomes routinised will be discussed in the book. Firstly, driving becomes routinised through car types; secondly, through the times and places in which driving occurs; and thirdly, through the age, gender and other characteristics of the driver themselves. The implications of outlining driving as a cultural practice in this way will be contrasted with approaches in social psychology that dominate road safety research, such as the reasoned action framework. The reasoning and wilful driver, central to psychological frameworks, will be contrasted with the desiring driver who is a participant in the social and cultural practice of driving. Driving is not, as might be expected however, routinised in terms of the laws of the road. Other factors have preceded and created the need for laws and indeed there is a play between the law and driving and its strong associations with outlaw activity that will be explored in later chapters.

Who do we become when we get into a car and take to the roads? Are we different people when we are behind the wheel, protected by the metal shell that enables us to insert ourselves into public domains from the comfort of our very own living room on wheels? What expectations do we have as drivers and how does the traffic and road system measure up to those expectations? How do we relate to the regulation of our driving by authorities and what ambiguities exist in our relationship to cars? What does being a driver demand of us?

The social and cultural factors involved in our attachment to the car and contributing to issues of road safety will be explored in this book. We can be largely unaware of these factors in considering cars and driving in traditional ways, since driving is an everyday practice that we take very much for granted. Nevertheless, the car shapes us in particular ways and makes demands of us as citizens through the driving cultures and car networks we are now immersed in (Sheller and Urry 2001). How the car shapes us and the particular demands that result from the practice of driving will be the focus of the book.

Driving is an activity that most people engage in, and strong feelings can often emerge when driving is discussed. While dealing with the daily traffic is a chore and often invokes frustration, driving can be a really enjoyable activity. The car is used for transport as well as facilitating pleasure activities, social lives and family connections. People also engage with cars in particular ways related to their gender, age and the kinds of desires they bring to the experience of cars. For some men, for example, cars are important to their expression of themselves as males, and driving large powerful or sporty cars is thus important to them.

Cars are approached by different people in different ways leading to different styles of driving. Some people stay in their place in the traffic queue while others will find ways to get around and forge ahead. Some drive with more attention on their precision in manoeuvring through the traffic, and others on the feeling of being in control of the car in other ways. Women have long been regarded as less able

Figure I.1   Highway traffic

drivers than men, however it is now the case that most women drive with equal capability. While men have prided themselves on their driving skill, they have more accidents and more serious accidents and this is related both to exposure and risk-taking behaviour (WHO 2004, 44 and 79). Women have tended to be more cautious and to be aware of the damage that a car can do (Scharff 1991).

Cars are seen as saying something significant about who we are. For many of us the smell and feel of a shiny new car gives us a feeling of being substantial in the world, of being out there. The style of the car is important to men and women, with many women not wanting to be seen in a car that suggests masculinity any more than many men want to be seen in a car that is small and powerless. Many people also prefer to be seen in a newer car rather than an old one, unless it is a special older car or they choose to live in an alternative, less consumer oriented world.

The car has facilitated growth in many countries and at the same time has coerced us into being particular kinds of people. It offers flexibility and equally compels us to live through it, to fit more in and do more with our time. The world we live in is fundamentally influenced by the presence of the car. Pedestrians are separated from traffic where possible, to facilitate the passage of cars and to keep vulnerable road users safe. Children are driven rather than cycling or walking to school because the roads are not safe enough for them (Thomsen 2005, Christensen and O'Brien 2003).

Cars have become more comfortable and easier to drive. They are more powerful and have better handling. At the same time, faster car travel has increased the pace of our lives. There are greater expectations of covering larger distances in a day and in less time. We can point to a myriad of gains from car use however there are also many 'side-effects' that we are aware of but reluctant to admit. The environmental effects are the most considered, the impact on our lives in other ways are less obvious.

Road safety research has a strong community appeal because it is expected that car travel should not be deadly and it is often considered that losing one's life or becoming maimed in the act of being mobile is 'accidental'. From the perspective of industry and governments crashes are unnecessary blemishes on the efficiency of the network and thus have significant economic as well as personal costs which are placed at the feet of individuals. Though funding of road safety research is at least in part, fuelled by the occasional community appeals to curb the 'unnecessary' carnage afforded by mobility through cars, at the same time a level of cost is accepted in the emphasis on the car in the lives of most people in the Western world. Yet millions of people are killed every year on the roads globally and increasing numbers of cars are being added to the roads each year.

Expectations of speed of travel through the motor vehicle have continued to climb. Cars have increased enormously in speed, power and comfort over the hundred or so years of their technological development. While authorities argue that an optimum speed of travel in motor vehicles in their current form has reached its maximum at speeds of 100–120km/h (ATSB 2003, 56), manufacturers continue to push for higher speed of travel in cars with speeds up to 300km/h being claimed even though there are few roads where such speeds could be achieved and the road toll is higher with higher speeds.

Blessing and curse – the double-edged car

The car brought a new experience of time and space to modern life, and allowed travel to places that it had not been possible to reach. However, it also reinforced the suburban sprawl that had long been a feature of many cities. Traffic made streets noisy and dangerous in ways they never had been before, resulting in the separation of traffic and pedestrians, with priority being given to motor vehicles. The sheer power of motor vehicles gave them the upper hand over the non-motorised.

Sheller and Urry (2000) argue that the car has been both flexible and coercive. It played a fundamental role in democratisation through the freedom of movement it facilitated permitting 'multiple socialities, of family life, community, leisure, the pleasures of movement' (743). However, it also coerces people to 'juggle tiny fragments of time in order to put together complex, fragile and contingent patterns of social life' (744). Communities have become synchronised to the rhythms of the roads and car networks producing a violence that interrupts and shapes the social environment. As Sheller and Urry note:

> Car travel rudely interrupts the taskscapes of others (pedestrians, children going to school, postmen, garbage collectors, farmers, animals and so on), whose daily routines are merely obstacles to the high speed traffic that cuts mercilessly through slower-moving pathways and dwellings. (744)

Yet the ambiguous nature of the development of the car and car systems, such as the convenience of faster travel and the consequent priority given to the car over pedestrians and other forms of transport, has been a feature of public discussion since the advent of the car. Ralph Nader (1965) declared the car 'unsafe at any speed' in 1965 and spurred the beginnings of an interest in safety. In 1972 Alisdair Aird published *The Automotive Nightmare* in which he clearly highlighted the benefits of cars in providing undreamed of freedom of movement, privacy and individuality while at the same time railing against the hypnotic power of the car and the industry which through its advertising encouraged people to ascribe to it 'a power of giving freedom which is so lyrical as to seem almost a drug' (7). Appleyard (1981) in turn argued that road traffic and its effects have been viewed by people from all social sectors as the most widespread urban problem.

Marsh and Collett (1986) claimed in *Driving Passion* that the car was 'an easy target for sociologists, economists, ecologists and the pundits and prophets of doom' with new books every year adding to 'the growing volume of abuse directed at the motoring world' (3). They maintained that the car had been grossly misunderstood since it was clearly much more than a means of transport and that it was 'the need to declare ourselves socially and individually' that kept car manufacturers in business in spite of the myriad problems created by the car (5). It is the psychological attachment to the car that makes the doomsayers wrong according to Marsh and Collett, and it is only when there truly is no need for the private car that it will disappear, which they cannot see happening any time soon. It is not only individual psychology that is responsible for the ongoing attachment to and dependence on the car however.

The emotional and passionate attachment to the car, it will be argued in this book, is a function of social and cultural factors through which individuals develop and cars are articulated or inscribed with meaning and made part of daily practice. The fact that we need to declare ourselves socially and individually through cars is a result of our particular culture rather than purely a function of the psychological nature of humans as Marsh and Collett suggest. The way in which freedom is understood and appealed to is also part of the culture of Western consumerism and capitalism which other cultures are prepared to adopt.

Freund and Martin (1993) considered the social dimensions of cars in *The Ecology of the Automobile* focusing particularly on the ideology of power and control enabled by the car.

> The cultural values of automobility find their psychic analogue in a masculine psychology of mastery and control, which embodies the culture of speed, power and the conquest of nature. (92–93)

Here the idea of social ideology is linked to and expressed through a psychology, whereas in a cultural approach culture is seen as providing 'framing' for the use and meanings of objects and practices related to them which are part of the background of everyday life (Couldry 2000). Freund and Martin nevertheless bring out the significant social factors contributing to car use, particularly the importance of power.

Figure I.2   City of roads. Aerial view of Sydney in the 1960s. Photo by Ken Redshaw

The taste for large powerful cars has continued to grow with the development in car technology, and manufacturers have used an appeal to increased power to induce consumers to buy the next model (Volti 2004). In America at the end of the first fuel crisis in the 1970s, 80% of cars had V8 engines (Volti 2004, 125). The tide subsequently began to turn in favour of smaller European and Japanese cars, however the preference for the 'muscle' car has not subsided with the more recent prevalence of SUVs and the re-emergence of the V8 engine even amidst rising fuel prices globally. Power and performance are the primary factors used in advertising, especially for large cars (Ferguson, Hardy and Williams 2003).

The priority given to the car and motorised transport in general is itself driven by the values of Western cultures. Graeme Davison (2004) in *Car Wars: How the Car Won our Hearts and Conquered our Cities,* gives a comprehensive account of the political manoeuvrings in Melbourne, Australia that brought about an emphasis on the car and car travel there. Davison documents the rise and rise of the car and outlines the lobbying by motoring groups which emphasised the ideology of freedom, and the dominance of government departments related to roads and motor vehicle regulation, over those dealing with public transport. Suburban development, it is clear, has made the car a necessary feature of modern life.

The car has had priority in planning and the real costs of cars have not been borne by individual motorists. The enormous cost of roads has been the burden of governments using public funds, notwithstanding the advent of toll roads. Then there are the emergency services and other costs associated with running a road system that are also borne by governments. Speed and efficiency with the promise of uninterrupted flow of traffic have been the focus of much urban planning through the separation of motorised and non-motorised means of mobility.

While values such as safety, quiet and individuality were emphasised over neighbourhood interaction in suburban developments, the progress of the car along with manufacturing processes meant even greater emphasis on speed and efficiency. By the 1970s the costs of both privatised mobility and privatised living, including isolation in suburbia, and related problems such as bored housewives and restless youth, began to be noted by researchers (Davison 2004). Where previous to the car children had access to 'a significantly larger geographical life-space' the car is now being used in modern society in order to extend that life-space which the car itself has restricted (Wolf 1996, 190). Children will then be keen to have access to their own car as soon as possible to make up for the restriction and dependence of the car dominated world.

More recently the tide has begun to turn. Where roadways were automatically regarded as the domain of motor vehicles there have been moves in Europe to enable broader social use of streets by creating 'shared' environments. Research in the United States on the effects of urban sprawl on crash statistics has shown that 'sprawl is a significant risk factor for traffic fatalities, especially for pedestrians' due mainly to the wide, long streets which tend to encourage excessive speed and the wide dispersal of homes, shops and workplaces (Ewing, Schrieber and Zegeer 2003, 1544). Another study in Wales looked at the 'sociability' of local areas and found that for 11 to 16 year olds living in areas with busy traffic the local area was perceived as less friendly and less safe (Mullan 2003). Mullan reported that 'children and young people were 100 times more likely to be killed as road users than by strangers' (352).

'Home Zones' developed in the United Kingdom as an alternative to the dominance of streets by motor vehicles, are designed to 'extend the benefits of slower traffic speeds within residential areas and give greater priority to non-motorised users' (Department for Transport, UK 2005, 5). Residential streets are designed to meet the needs of pedestrians, cyclists and local residents and to provide safer streets where children can play. The aim of the Home Zones has been to reduce traffic speeds thus improving road safety to improve the quality of the environment and increase play and other activities in the street area through community involvement. Besides these benefits other effects have been achieved such as reducing crime in the area and the development of stronger, more integrated communities (81).

Cars as neutral technology

Despite the history of debate about cars and their benefits and costs, it is individual road users who have been held accountable for road carnage and the emphasis in road safety has been on assisting drivers to be 'error-free' (Featherstone 2005, 4). Technological developments have reduced the impact of the car on the environment and improved handling, however, cars have had an enormous impact on the way communities function and how we live our lives. In all of this the car is implicated in important ways in shaping who we are. The car is part of a network of mobility that has, along with other factors, fundamentally changed the way we live.

The psychological research that is central in road safety tends overwhelmingly to focus on the individual without really drawing on individual experience or attempting to relate individual ideals back to broader cultural ideals. For example, there is a focus on those who rate high on sensation–seeking or adventuresome scales without questioning where the emphasis on using cars to fulfil a desire for adventure comes from or is reinforced (Dahlen, Martin, Ragan and Kuhlman 2005). In other words, road safety research does not confront mobility as a system in that it does not connect the operating social norms to broader issues that come back to the major stakeholders such as manufacturers and motoring organisations, governments and cultural values.

One of the reasons for the focus on the driver, and not the car, is that historically, cars have largely been regarded as neutral in that they are seen as dangerous only when they are not handled properly, rather than being dangerous in themselves (Jain 2004). Given the enormous significance attached to cars beyond their practical significance, their 'neutrality' has to be regarded with some scepticism. In addition, the sheer weight and 'bullying' power of a motor vehicle compared to a pedestrian or cyclist evokes a power that is not available to those who are not motorised, suggesting that with the growth of car use, pedestrians and others were simply bullied out of the way.

Jain's analysis (2004) has explored the legal processes that brought about an emphasis on the fault of pedestrians and drivers that meant manufacturers were not held accountable for the design of vehicles. She points out that SUVs are openly marketed as providing protection to their occupants at the expense of others in smaller vehicles. Jain's analysis of the legal battles over culpability for injuries incurred by the 'ornamental protuberances' (78) such as tail fins on cars for example, shows how children's needs were dismissed in legal thinking about car design and the blame was often placed with mothers for not keeping children off the streets, where they had been used to playing, and out of the way of cars.

The car being motorised and thus lacking a 'mind of its own' has been seen as the vehicle of rationality and consequently any part in the development of current social formations has been denied. The rationality of cars and car systems is part of the imagined effects and consequences of technology. The promises applied to the car are not unlike those being offered by Bill Gates for the smart house: ' ... the networked house is an emblem of millennarian harmony, an essential part of a post-industrial society of revitalised individual freedom and autonomy where, equipped with home interactive telematics, the private individual can herself set the terms of engagement with an external world while simultaneously maintaining secure boundaries around private space' (Allon 2004, 266). The dream of free-flowing movement, unlimited flexibility and connection has been applied to the car and is constantly reinforced in current advertising (Lupton 1999). The technology is considered neutral, manipulable and controllable and yet it has had a profound effect upon social formations.

Sociology too has regarded cars as neutral technology and yet the consequences of the car and its logic have been enormous through the reconfiguring of civil society and people themselves (Sheller and Urry 2000). The car has had many unintended consequences. Pollution and overcrowding problems and planning implications – women isolated in suburbia, the increased distances of commuting, the budget for road building and maintenance – along with the frustrations and demands on the

Figure I.3　　Bikes, traffic, people in Barcelona. Wide streets for movement of motor vehicles is often combined with narrow footpaths for pedestrians in this case shared with space for parking motorbikes

individual of the convenience of cars, were not envisaged in the early days. Not only were the limitations not imagined in the promises of enhanced freedom, but the car as machine was considered neutral in its effects.

The limitations of the logic of the car and car systems include the reduction of choice through the reconfiguring of society in the physical separation of home and workplace, home and commercial centres, and a shortage of time. The car becomes necessary to the extent that alternatives such as walking, cycling, bus and rail appear inflexible and inconvenient (Sheller and Urry 2000). The systems of organisation, rationality and logic presupposed by the car in turn produce resistance. Sheller and Urry cite local protests over the expansion of major highways and objections to expanding road networks by environmentalists. We could also note the infringement of rules by drivers as a resistance to the logic of the system where both car and driver become the irrational resisters, evading the inevitable logic of the system.

One simple example of the influence of the car on our way of life is in the loss of the front porch or balcony as social space connected with the street. Where people used to sit on their front porch and observe and talk to passersby the garage or carport has now become a prominent feature of urban architecture (Volti 2004, 111). The garage had previously been located at the back of properties as cars were noisy and dirty, now it is not uncommon to see two and three garage doors in front of houses with connecting doors from the garage directly into the house. The car is the mobile part of the house extending 'home' beyond the boundaries of the house. It has become part of our lives to the point where it is brought into our homes. At the same time the street has become the domain of the car rather than a site of sociality with the garage intruding between home and street. One steps from the car into the garage and then directly into the lounge room of the house bypassing the street entirely.

There are important ways in which the car has been shaped by gender and in turn shapes and characterises gender. John Sloop (2005) has looked at performativity and gender in relation to cars, and his analysis draws out the contradiction between the apparent gender neutrality of the car and the gender inscription of the driver. There is no doubt that in significant and complex ways gender and cars are connected. Not only are cars genderised but the relationship between the car and the gender of the driver has important cultural and social implications that will be explored extensively in Chapter 3.

The extent to which the car has shaped and is shaped by social action and interaction has become the focus of commentary on cars more recently, particularly in sociology (Wright and Curtis 2005, Dant 2004, Sheller and Urry 2000) Dant in particular has explored the assemblage of driver-car as a form of social being that produces a range of social actions associated with the car (61). Through the idea of 'affordance' he has examined the car as offering both mobility and motility, that is, spontaneous and independent movement, in a progression from horse transport. As an affordance, the car has a real physical resistance that has shaped and is shaped by human action and is given meaning in a variety of forms related to complex features of the cultures it is embedded within. As a social object, the car is not fixed or pregiven in its meaning – there are substantial symbolic elements that are essential to the constitution of the car (Urry 2000, 65).

Not only has the car had a place in changing social configurations it has also played a part in the fundamental constitution of citizens. Actor Network Theory following Latour (1993) has recognised and theorised the interplay between humans and technology showing important cultural as well as embodied interconnections. Mike Michael (1998) argues for the influence of the car itself and a hybrid of human and car he refers to as a 'cason' and the impact of this hybrid in phenomena such as road rage. Road rage, he argues, is a result of a dynamic where 'loss of social control' results from greater technological control or hyperhybridisation in which the human is obscured within technology (Michael 1998, 133). In this hybrid of car and person, there is a move to 'the machinic conditions of skilful technological conduct', determined more in this situation by the car. The human, beyond social control is 'overcome' by the car. For Michael the 'machinic conditions' determine the person. The promises and expectations of the car lead to a hyperhybridisation in social and interpersonal relations to the car. Such an extensive investment in the car as part of the self leads to aggressive responses to infringements of the car's space and dignity.

The purported neutrality of the car then, is not so neutral but shapes and creates forms of organisation and human interaction, as many have recognised (Latour 1993, Michael 1998, Urry 2000). Humans are changed in relation to cars, and cars are shaped in relation to human needs and comforts and those of a broader system. There is no doubt that the car is greatly loved and valued in very personal ways (Marsh and Collett 1986). At the same time, however, we have been unwittingly shaped in relation to the car and a car dominated society. It is thus important to consider the car and driver together in order to better understand the impact of the car and attachments to it, rather than simply viewing each separately and emphasising one at the expense of the other.

Michael (2001) argues that the car is invisibilised in the tendency to focus on the failings of human nature. The assumed neutrality and adaptability of the car is up against the limits of human nature. But the human is made fixed and inflexible to some extent in these ideas of human nature themselves. References to systems of control within which humans are moulded to the machine and controlled within the systems it entails are based on a confrontation between social and individual control where the individual is limited by their own 'human nature'. At the same time, very productive cooperation generally operates within car systems enabling the whole functioning of the system. The discourses Michael discusses, the Royal Automobile Club and the Automobile Association, resort to 'human nature' in their focus on the aberrant behaviours that infringe the rules of legality and etiquette. Human nature in this sense is contrasted with the car-made civilised person.

The efficiencies of car systems and the cooperation required by regulation enable mobility, but the expectations of the rationality of the machine extend far beyond the reality. The system is then perceived as controlling and limiting in itself. The promise that control of the car will be empowering is way beyond the actuality in emphasising the efficiency and neutrality of technology shaped by the empowered controller.

The car embodies diverse expectations, comforts, fears and hopes promising seemingly unlimited control. At the same time it is an obstacle in itself to that control. The car can only be controlled and used in limited ways. It cannot fulfil in reality the needs and desires for unfettered freedom and control that it is assumed to embody. Cars were originally regarded as dangerous machines but as they became an increasingly necessary part of everyday life, the blame was removed from the technology of cars themselves and shifted to individuals, first pedestrians, then women (Jain 2004). Rather than the technology being removed or modified, or the manufacturers held accountable for the damage cars could do then, attempts were made to exclude 'problem individuals' from using it.

While there have been significant changes in car design this perspective is still very much at the centre of thinking in road safety and denies the implicit role of broader cultural factors, including the car itself and the meanings that are part of it. Thus there has been an emphasis on the car in some areas and in other areas the focus has been on the driver, however, in order to confront the impact of the car and our attachment to it, we need to deal with both together and the interaction between them.

Attachments to the car have not been adequately recognised in the road safety literature and this has meant that solutions have tended to be technical rather than social. At the same time the idea of human functioning underlying theories of driver behaviour have relied on an opposition between will and desire that will be examined in more detail in Chapter 4. Social psychological research in road safety has been based on the theory of planned behaviour emphasising identification of behavioural characteristics associated with risk and the tendency to commit violations (Parker, Manstead and Stradling 1995). The theory reveals many important elements of behaviour that are involved in driving accidents and committing infringements, such as feelings of invulnerability, illusion of control, lack of thoroughness and failure of observation (Parker, Reason, Stradling and Manstead 1995). The analysis, however, is limited to individual intentions and does not extend to the culture of driving as a whole and the expectations and ideals that are operating there.

Some behaviours, such as speeding, may be implicitly condoned in the broader social and cultural context. Wilful lack of concern for one's own and others safety noted in this research may reflect the type of moral norms associated with driving in the community generally, and the relative weakness of norms assumed to be associated with driving, such as a priority for safety over speed. In addition the emphasis on wilfulness and the need to control desires has a history (Valverde 1998) that is implicated in the way the relationship between cars and drivers is constructed. Freedom has involved in Western societies the ability to manage desires through the will (Valverde 1998, 17). The desires that must be managed are in turn created in the context of social identifications and attachments.

Road safety research, it must be noted, does not place a great deal of emphasis on the themes in advertising noted by cultural and social theorists, themes such as the significance of power, control, individuality and freedom, as these are portrayed in car advertising. Social researchers have noted that car manufacturers and advertisers have exploited and enhanced the symbolic dimensions of the car and have explored the symbolic significance of mobility extending beyond the practical functionality of cars (Featherstone 2005). The affective and instrumental appeal of the car, the various dimensions of the car including styling, and the symbolic and affective motives for car use have been investigated, contributing to a broader understanding of the meanings of the car (Anable and Gatersleben 2005, Wright and Curtis 2005, Steg 2004).

Important moves have been made more recently to make road safety a public health issue so that there is more sharing of the responsibility for car travel. Featherstone (2005) discusses the WHO (2004) World Report on Traffic Injury Prevention that seeks to redefine traffic 'accidents' as a public health issue so that the traffic system is held accountable for failings resulting in death and injury and not just individual drivers. Importantly car travel can then be looked at in relation to the costs of different modes of travel.

As governments continue to put money into faster more efficient roads, often at the expense of public transport, and as manufacturers, through their advertising continue to promote cars as the vehicle, not only of freedom, but of radical excitement, individual expression and masculine display, without any acknowledgement of the potential costs, young people who are over represented in the crash statistics, can hardly be blamed for not realising the consequences of their enthusiastic embrace of the motor vehicle. In this sense the system is failing young people and then blaming the victim. There are myriad reasons why young men in particular approach cars and driving as they do. Expressing and displaying masculinity through cars has been encouraged and considered appropriate despite the danger presented by cars. Cars and displays of car handling skill are also an expression of self-respect and self-esteem, especially for those from lower socio-economic backgrounds for whom the power of the car presents a major avenue through which to gain self-respect and social kudos (Walker 1998).

Social and cultural perspectives on driving

Driving is cultural in the sense that it is something that people engage in in a range of ways particular to other cultural factors such as gender, status or socio-economic background, geographic location and values and ideals. In the research on which this

Figure I.4   Faster roads, faster cars, fast moves in the traffic

book draws driving is seen as revealing important aspects of an everyday practice that is not just individual but also imbued with cultural meanings. Driving is an activity that would not normally be regarded as expressive of culture. Yet it is an everyday practice, a lived experience characteristic of the modern world that is largely taken for granted. How it is lived by different people and how those experiences overlap with those of others is of interest from a cultural perspective. Exploring driving as a social and cultural practice thus involves seeking to discover the experience of living with cars in all its 'complexity and ambiguity' (Sheller 2005).

Amongst the literature that has grown up with the car there has been some recent examination of car cultures where the focus is on types and uses of vehicles within different 'subcultures' (Miller 2001). In contrast to *car* cultures, the central concern of the research outlined in this book is with *driving* cultures, that is, with the ways in which cars are meaningful, and the particular driving styles through which relations to cars and to the roads and traffic are expressed. Cars are inscribed with particular meanings that are part of the very 'metal' of the machine, or 'baked into' them (Ormrod 1994). The concept of meaning being 'baked into' things has become a standard in Actor Network Theory originating with Ormrod (1994) for whom the microwave oven as a cooking device has meanings burned into it through its design, manufacture and marketing.

Cars have always *meant* freedom, independence and convenience, as well as status and sex appeal. These meanings extend far beyond the merely instrumental function of the car as a means of transport. Nevertheless objects are not fixed in their meanings and their meanings are not preformed or given (Urry 2000, 65). Cars mean different things to different people and can be used in a myriad of ways not necessarily foreseen by manufacturers, according to social and cultural configurations.

Social and cultural research relating to a practice such as driving then, is not only concerned with the systems of mobility such as road and traffic systems, manufacturing and government processes. It is also concerned with the relations that constitute that system, that is, relations between cars and people and between people in different cars, as well as between drivers and the authorities that regulate driving, and the discourses that surround driving in various contexts from magazines to road safety literature, manufacturers and advertisers. The extensive symbolism of the car reaches far beyond the facilitation of mobility; it is a habitat, a second skin where passengers are enclosed in a self-contained world of their own. Drivers become increasingly isolated from social restraints through the retreat into a private world where demands from others are intrusive (Graves-Brown 1997).

The car has been part of a wider cultural process – the increasing privatisation and individualisation of experience. As the market provides an increasing plethora of goods we choose according to our individual tastes and desires what we want to listen to, what we want to see/watch, where we want to go and what we want to eat (Graves-Brown 1997). The mobile and home centred way of life in Western societies has been referred to as 'mobile privatisation' by Raymond Williams in order to denote the changing social formations occurring with the advent of radio sets, motor cars and other home centred electrical appliances (Williams 1975, 2).

Margaret Morse situates television as a cultural form within the larger socio-cultural context of everyday life by relating television, the freeway and the mall through a 'fiction effect' she refers to as distraction (Morse 1990, 193–194). This is a 'partial loss with the here and now' and related to split belief in which one momentarily sinks into another world knowing it is not real. A level of unreality involved in the freeway, the mall and television invokes feelings of both pleasure and boredom, in which each is considered 'a "vast wasteland" and a waste of time as well as a devotion allied with the American dream' (Morse 1990, 194). The separation of home and work and the need to move between the two, usually in private cars, has contributed to the sense of distraction as a defining characteristic of 'mobile privatisation'

Others such as Tim Edensor (2003) have argued for the emergence of new forms of sensuality in the interaction between car and driver and the particular bodily engagement that produces a '"feel" for the road' (159). A variety of experiences are played out and potentially conflict with each other on the roads. Car enthusiasts have asserted a dominance due to their claims to a more authentic experience of driving as opposed to tourists, who can be seen as passive and disembodied, and present a nuisance to those who appreciate speed (Larson 2005). Much of the argument by car enthusiasts has centered on the embodied experience of the vehicle and the road that enthusiasts see themselves as able to master. Other drivers are also embodied in their experiences of driving, however, though they may have quite different experiences of that embodiment. For those who live in a particular locality a different sense of ownership and entitlement may apply to the roads in the area. Their familiarity with the nuances of the road could give added confidence to their ability to handle it and add to their impatience with those who meander along it merely 'looking' or unsure of where they are going.

Figure I.5   Different traffic needs compete in the city. Chinatown, New York

De Certeau's 'Walking in the City' (1984), famous for its phenomenological enunciation of walking, opened up possibilities for recognising and addressing the 'operational logic of culture' (Thrift 2005). Concerned with everyday experiences, cultural studies has to some extent been involved in exposing what is 'evasive' and 'hidden', the operations of 'culture' of which we may, in our everyday activities, be unaware. Providing a language for things we do and the ways we do them is to 'enunciate' and to give representation to (Morris 1998, 110).

De Certeau has been criticised by Nigel Thrift (2005) for his privileging of 'walking in the city' as a more authentic practice than driving a car or walking from a car, when much walking, he points out, is *derived* from car travel. For Thrift, however, de Certeau can also take us beyond the representational analysis of the car as symbolic manifestation of various desires, to consider the embodied practices of driving and being a passenger, produced by the system of automobility. Thrift describes the ways in which the car and the driver have become melded to each other through technological developments, moving to the point that 'bare life is being laid bare', while new forms of embodiment are being developed.

Drawing on the lived experiences of people engaged with cars, traffic and systems of automobility then, is important in understanding where improvements could be made towards saving lives, inequalities could be addressed and resources could be made to go further and last longer. Cultural research provides some important approaches for unearthing and expressing the significance of the car and our attachments to it. In the following I will outline some of the key concepts to be employed in the book.

Articulations of the car

One of the concepts that will be important in exploring the kinds of social and cultural relations generated in and through cars is the idea of articulation (Redshaw 2007). Driving cultures have emerged which can be related to particular dominant articulations of the car whereby cars are appropriated into dominant meaning systems such as that of aggressive individualism.

Driving cultures will be considered as articulations of the car in the sense that Lawrence Grossberg (1992) used the concept in his book on rock music culture. Grossberg was aiming to provide an alternative to cultural studies that equated culture with communication, in order to 'describe the complexity of effects and relations circulating through and around culture' (1992, 45). He was concerned with 'particular configurations of practices, how they produce effects and how such effects are organized and deployed' (1992, 45).

> Articulation is a continuous struggle to reposition practices within a shifting field of forces, to redefine the possibilities of life by redefining the field of relations – the context – within which a practice is located. (1992, 54)

By applying the idea of articulation to driving cultures, the intention is to encompass, not only discourses relating to cars and driving, but also the implications of the car itself, and the associations and attachments to cars expressed in discourse. The affective appeal of the car, in addition to its function in facilitating mobility, has enabled the car to become a means of social and individual expression (Anable and Gatersleben 2005). Nevertheless, particular articulations of cars have dominated the social and individual meanings inscribed in them. The meaningful associations that have shaped car use include aspects of the context that are taken for granted, such as the connection between the development of racing car technology and road vehicles which has brought with it implications for the ways in which cars are driven, including an emphasis on speed and power. Racing associations have shaped and articulated the context of car use and in particular, emphasised the aggressive assertion of cars at the expense of other forms of mobility, while at the same time downplaying the destructiveness of the car.

Social framing

Articulations of cars then, are connected to actual experiences of cars and the pleasures they afford, but also to a broader context of powerful interests. Manufacturers vie for position in the market place by appealing to the fantasy and pleasures of the car and in the process are largely involved in framing articulations of cars. Car advertising frames the practice of driving in particular ways, playing a part in the shaping of expectations about cars and the experiences they are able to induce. Framing here has a special sense connected with mass media:

> … in a society of almost universal television consumption, and largely shared patterns of programme availability, the simple fact that television is 'by definition … common to all' itself grounds its function as a frame for the social. (Couldry 2000, 14)

It is argued here that car advertising is an unquestioned part of the background of daily life that is not necessarily reflected upon, but which nevertheless leaves an impression:

> A media culture has emerged in which images, sounds and spectacles help produce the fabric of everyday life, dominating leisure time, shaping political views and social behavior, and providing the materials out of which people forge their very identities. (Kellner 1995, 1)

The media, Kellner argues, provide models of 'what it means to be male or female, successful or a failure, powerful or powerless':

> In a mass mediated culture, it is representations that help constitute an individual's view of the world, sense of personal identity and gender, playing out of style and lifestyle, and socio-political thought and action. (60)

The 'immersion' theory of media at least allows for some acknowledgement that advertising has a part to play in the meanings that are applied to cars and the practice of driving. The relationship between advertising and behaviour is not one-way (Morphett and Sofoulis 2005) as advertising draws on and reinforces already existing articulations of cars, such as those drawing on racing and rally driving themes. However, connections can be shown between dominant themes of advertising, such as the forms of masculinity, individualism and competition that they draw on, and the interpretations of those themes, as well as what is considered suitable behaviour on the roads.

Driving as a cultural practice

The focus in this book is on driving as a practice and the 'cultural orders', flows, inequalities and asymmetries that are involved in 'culture' as 'the complex and uneven working out of power' (Couldry 2000, 110). In other words driving culture is not a unified culture but made up of a range of disparate and complex experiences. However, within the irreducible individual differences that comprise driving culture, there are wider forces influencing that experience, including cultural formations such as gender, age, wealth and location. Cultural forces are not only determining conditions, they are also *limiting* conditions, imposing limits on the type of experience that is possible in and through cars depending on who and where one is, amongst other conditions (Couldry 2000, 51). *Driving* culture is thus concerned with driving practices in a particular sense. Driving practice encompasses the discourses and embodied, emotional and desiring dimensions of driving, but it also examines the routinisation of driving, which makes it a cultural practice. There are likely to be a range of driving practices but these are nevertheless limited by the social and cultural engagements with the car that are allowed and facilitated by car use, as well as the car itself.

Figure I.6   **Speed and safety discourses have been around since the advent of cars**

Driving as a practice can thus be outlined by drawing on some of the dimensions through which it becomes routinised. There are three important dimensions that will be outlined in subsequent chapters: firstly, driving as a practice becomes routinised in relation to particular cars, which are experienced in particular ways; secondly, it is routinised through particular embodied experiences of driving related to where in the complex system of automobility they occur, and thirdly, through various demographic characteristics of the driver. In what follows I will draw on focus group comments and discussion to outline these dimensions. The dimensions of driving practice as I have identified them are interrelated, as will be evident in the statements that follow, and are certainly not exhaustive. They merely help to flag the sense in which driving, although a highly individual experience, is routinised as a practice.

The authentic driving experience has been considered to belong to the car enthusiast or those who cherish the experience of speed and controlling it, and challenge their driving skill on demanding roads. This is usually men who are concerned with the power and speed of the vehicle and its engine and body specifications, handling ability and performance. In considering driving as a practice, although these experiences are certainly lived differently, one type of driving experience is not considered here as more authentic than another.

Considering driving as a cultural practice and drawing on the lived experiences of people with cars helps to give some substance to the relationships between cars and those who engage with them, and to address the dimensions on which the car is implicitly meaningful in those relationships, as well as how it shapes other relationships such as those between young men and women.

Driving as a practice crosses a range of other practices in which people engage and which are facilitated by driving. Driving is a practice in itself, and is articulated by a range of practices that are separate but interlinked, including practices of the media, consumption practices, regulatory practices and public/private interaction (Couldry 2004). The specificity of this practice and the ways in which it could be different and for whom, is the political and ethical dimension of the research. As a practice, driving encompasses symbolic dimensions that are bodily and mental, emotional and instrumental. These are routinised in relation to driving and thus are social (Reckwitz 2004). Driving cultures then, embrace this range of practices and amalgamate them into driving practices and as cultural research focuses on the meaningful and everyday relations and significance of the car, and the practice of driving.

Bringing cultural studies to this field and applying cultural approaches can draw out areas within the practice of driving that are less visible because they are so implicit a part of everyday practice, and imbued in the practice of driving within the context of broader cultural practices of consumption, of time and place, as well as things, that they are rarely subject to investigation and critique. Examining driving as a cultural practice shows that other aspects of culture, such as the role of gender in driving practice, are related not only to implicit meanings that are taken for granted, they are also part of the power structure that cars are implicitly involved in.

Examining dominant cultures and highlighting implicit ways in which driving practices function, requires the engagement of different perspectives. Gender is a paramount issue in driving research since men are particularly over–represented in statistics and driving practice could be said to be male dominated. All things to do with cars have long been male dominated and male centred, though women have also been involved from the invention of cars, even if in significantly smaller numbers. However, 'masculine' desires have considerably dominated the prevalence and invasive presence of cars. Not only is understanding the perspective of men important

Figure I.7   Diverse literatures contribute to the car cult

in this context, giving voice to other perspectives that are often downplayed, such as those of women, the elderly and young people when it comes to cars and driving practice, is also important. Other forms of mobility must also be taken into account – pedestrians, cyclists, motorised and nonmotorised wheel chairs as these perspectives are important in shaping the future of mobility and quality of life.

Mobility will be considered from a broad cultural approach taking into account as many different perspectives on mobility as possible though the book is predominantly concerned with the dominance of the car and experiences relating to it. A range of research approaches will be drawn upon hopefully keeping in perspective the limits of each. Many diverse research perspectives contribute to the overall picture but here it is not a picture of road safety that I am interested in but of driving cultures and the various practices that make them up, including research practices.

Young driver research

The book draws on research from the Driving Cultures program of research, particularly the Transforming Drivers project, a partnership with New South Wales motoring organisation NRMA Motoring and Services. Focus group studies were carried out in New South Wales between 2000 and 2005 with young drivers aged 17–25 years in city, urban and country areas. Some of the studies have been reported in other publications and major reports are available online.

Although the major global road safety statistic is amongst pedestrians in less developed countries where the car is currently being taken up with vigour, the focus in Western countries is often on young drivers. Young drivers are over represented in the statistics in most Western countries and are seen as a major area of concern where a significant impact could be made on the national road statistics if changes could be made.

There has been a substantial amount of research on young drivers in recent decades. Most of this research is quantitative and conducted from an epidemiological or psychological perspective concerned mainly with statistical generalisations. The aim of such research is to get a broad and depersonalised picture of the young driver 'problem' and what it is about young drivers that makes them more prone to crashes. It concentrates primarily on aspects of driver performance such as offences, crashes and errors (Parker, Manstead, Stradling, Reason and Baxter 1992; Parker, Reason et al.1995), driving exposure such as night time and weekend driving with passengers, and constructs such as attitude and sensation (Jonah 1997) and adventure seeking (Berg 1994). The recognition of attitudinal and motivational factors has led to increasing research on lifestyle factors to include factors outside driving itself that could be contributing to young driver behaviour (Moller 2004).

The involvement of social science in investigating broader social factors has meant use of qualitative research methods to provide more in depth and specific information, though research in the road safety field is still predominantly quantitative and expected to follow quantitative formulations. Another aspect of the bulk of the research that is worthy of note is the focus on accident data rather than everyday driving (Ranney 1994). Taking into account lifestyle factors goes some way toward addressing this but there is still less focus on the wider contributing factors themselves than on individual behaviour. Social and cultural factors need to be considered much more intensely so that the significance of their role can be factored into education and planning. While social norms are structured into the theory of planned behaviour for example, the specific norms and how they function has not been intensively investigated. Social and cultural theory has a significant contribution to make in this regard and it is with this in mind that the research projects drawn on in this book were undertaken.

Two major studies were conducted as part of the Transforming Drivers research. The first was a study of car advertising with a particular focus on television ads. Sixty young people participated in nine focus groups across Western Sydney, Inner Sydney, and the greater West of New South Wales. The groups invited wide-ranging discussion on car and safety advertisements. The focus groups had the principal aims of uncovering some of the complexities in how meanings of driving and cars are produced in response to media, especially among youth audiences, and to test how viewing and commenting on television car and safety ads worked as a way of opening up discussion about the social values, cultural meanings, personal identities, risky practices, and various experiences associated with cars and driving by the under 25 age group. Of the total of 60 participants, 30 were male and 30 were female and ages ranged from 16 to 25 years and is extensively reported in Sofoulis, Noble and Redshaw (2005).

The second was the Gender, Ethnicity and Control study. Ten focus groups were conducted in various locations in New South Wales including Western Sydney, Wollongong, Goulburn and Wagga Wagga with a range of ethnic groups. There were a total of 65 participants ranging in age from 18 years to 24 years and 30 were men, while 35 were women. The aim of the study was to look at the different ways in which driving practices are experienced and performed as well as the ways in which cars are enabling, according to gender. The study is reported in detail in Redshaw and Noble (2006). Four participants had learner licences, 38 had provisional licences and 19 had their full licences. Two had been disqualified and two did not have a licence. Focus groups were used in order to draw on the social interaction involved in discussion of cars and their centrality for the young people. Focus groups were tape recorded and transcribed and then coded according to key themes using NVIVO software.

Comments from focus group discussions are included as they relate to the themes of the book. They have been selected for their relevance to the discussion with careful attention to maintaining the original meaning of the comments in the context of the focus group discussion. The themes that are discussed in the book have been developed through the course of the research and so are connected to it and do not merely attempt to shape the research to the themes of the book. Every attempt has been made to situate the themes and research in relation to one another while highlighting less obvious points and those that are not often considered in other types of research. Detailed discussions such as those that occurred in focus groups offer the possibility of looking in greater depth at the major themes and of considering new and unexpected themes that emerge from those discussions.

Social dimensions

A major theme in the book will be the social and cultural nature of driving. Even though in Western cultures cars are considered the ultimate vehicle of freedom and individuality, a great deal of social cooperation is required to make this possible. Social cooperation occurs in all countries in relation to mobility, even the chaotic streets of Bangkok, Delhi and Peking involve enormous amounts of cooperation. The addition of regulation can aid the success of that cooperation, though regulation often has the stigma of limitation of individual freedoms attached to it.

Social cooperation has also brought about the dominance of the car, which through its sheer bullying power has been able to take over the streets from pedestrians and children and other non-motorised forms of transport. The ability of the car to go anywhere has been facilitated by the building of roads, driveways and parking, at enormous cost to the community, a cost which is not often factored into the transport of goods but is borne by the public purse.

The increased emphasis on the individual and individual desires and needs has created an atmosphere of competition on the roads that belies the social cooperation implicit in it. The book will discuss the articulations of cars that encourage aggressivity and represent the streets as combat zones where it is a case of everyone for themselves at the expense of interaction that is supportive and friendly. It is common for people to regard other drivers as 'idiots' one has to be wary of and to respond aggressively to mistakes or infringements of one's passage or space on the roads resulting in the idea of 'road rage'. The social implications of this buildup of discontent will be discussed. The sense in which the car promises more than it can deliver, particularly as populations increase and intensify is explored throughout the book.

Personal mobility through the individual car in its current form is unsustainable for a variety of reasons. Firstly there are the enormous environmental concerns and the limits on the future of oil availability to say nothing of the overcrowding of the streets with increasing car numbers. Further to this however, there is the social unsustainability of the spiralling aggressivity and continued promises of increased speed and performance of cars. The car issue does need to be considered as a community concern since access to mobility in whatever forms is a social and community responsibility not just an individual problem. We cannot simply ban older people from driving without considering how they are going to be transported for example. Community services have begun to develop for transporting older people from their homes to community facilities. Mobility in this community sense that is neither private nor mass transport is likely to increase in the future along with possibilities for car sharing and alternative powered vehicles.

The book concentrates on intensive consideration of the social and cultural implications of driving and cars. The first chapter will examine the type of car as a cultural factor focusing on the interplay between driver and car and argue that cars are articulated to express particular characteristics that impact on how they are driven. The second chapter is concerned with experiences of driving in relation to time and place and how these are routinised affecting the way in which different cars are driven and different drivers are likely to drive with a particular focus on experiences of boredom and pleasure. The third chapter focuses on how characteristics of the driver, primarily gender and age, impact on the style of driving and employs a notion of 'combustion masculinity' to highlight the dominant style of driving and how young men relate to it. Chapter 4 examines the framing of desires and emotions in driving and offers a new way of formulating desire and the relation between reason and emotion that is an alternative to the overwhelming focus on the taming of desires created through. In Chapter 5 the car and governance are discussed including consideration of the history of road rules and the influence of the motoring lobby, the role of the police and ideas of self-control. Chapter 6 offers some alternative ways of thinking about mobility into the future to bring together the main points of the book by discussing alternative future forms of mobility, challenging the 'need for speed', outlining 'hydraulic masculinity' as an alternative to 'combustion masculinity', and employing a notion of hospitality as an ethics of the road that can highlight cooperation and generosity to replace the emphasis on competitive aggression as a central theme of cars and driving. It is hoped that the investigations presented here will make an important contribution to safe and accessible mobility in the future.

Chapter 1

Enticing Cars and Driving Styles

The first dimension of driving as a cultural practice that will be explored in this book is the car. Particular cars have reputations for 'demanding' to be driven in particular ways. This is not just popular myth however. There is some substance to the idea that 'hot cars' demand to be driven in ways that could be described as 'hot'. At the very least, the type of car suggests a style of driving whether that style is expressed or not. Powerful cars 'suggest' take-off power though this may not always be used. Young people's comments on their driving styles in relation to different types of cars and examples of car advertising inform and illustrate this theme showing for example, that a large car with sports pack such as a Ford Falcon XR6 produces a driving style that may otherwise be alien to the person driving:

> My dad has an XR6, right, and learning to drive on an XR6 was quite difficult because as soon as you hit the accelerator you're doing 50 K's already … I get the worst road rage in that but I think it's because I know that I can beat that guy who's driving a little slow car, so why not? (Female 3, Lawson)

How cars are articulated in advertising and on the roads informs people's relation to different car types. The Ford Falcon XR6 is built and promoted as a mean, aggressive car that has to be driven accordingly. Aggressive power and speed have been 'baked into' it.

The first dimension of driving as a cultural practice then is the car itself and its variations, and how articulations of particular cars contribute to the driving cultures related to them. While improvement in technologies can be achieved by focusing on the car in isolation from the driver and the road environment, there are other aspects of cars that are equally important, in particular the meanings that are 'baked into' cars, from the factory floor to the public roads and the owner's garage via the framing of the car through the media. The expressions given to cars on the road are informed by these framings. The ways in which cars will be driven and by whom then, are influenced by the social framing of the particular car, and this is built into the car. Further, it will be argued in this chapter that cars embody particular cultural attitudes and thus that attitudes extend beyond individuals.

Cars and drivers are implicitly connected in important ways that will be highlighted in this chapter. The car can become something different according to who it is being driven by, even though car and driver have their own separate and particular meanings. The driver themselves can likewise become someone different depending on the car they are driving. Some will need to drive differently in order to maintain their identities, such as men in small cars who might drive more aggressively rather than according to the expectations of how one drives a small car. Women similarly might find themselves driving more aggressively in a larger, more powerful car.

**Figure 1.1   Ford Falcon is a popular model in Australia and this XR6 with
                sports pack is attractive to young men**

The conduct of cars

The relationship between cars and drivers has been theorised in a variety of ways
as we saw in the previous chapter. Although in legal terms drivers have been held
accountable, in the early days this was not necessarily so (Jain 2004) and drivers
themselves have variously placed the blame with cars and roads. Cars are sometimes
'blamed' for the failings of drivers such as when speed is attributed to the car
'just wanting to go' and when cars just 'run off the road' because of their power.
Sometimes control is vested with the car but ultimately it is vested in the driver. A
Ferrari might be promoted as a car that demands 'high driving skill' (Pollard 1974,
annotated illustration opposite page 25 by John Carnemolla) but it is also the type
of car that offers the greatest potential for losing control as if it might be capable
of overpowering the driver despite their superior driving skill. This ambiguity is a
theme that runs throughout the history of the car and is perhaps one of the reasons
that the future of cars is seen to rest primarily with improvements in technology such
that the car more or less drives itself in order to overcome driver error and remove
the temptation to take risks. The contest between driver and car highlights the
ambiguity of the relationship and the ways in which cars are articulated. Exploring
the articulations of cars then will involve some elucidation of the relationships
between cars and drivers, which are of course many and varied.

Cars are the ultimate symbol of freedom, independence and individualism. They
offer the freedom to 'go anywhere', whenever it suits and with whom one chooses in

the privacy and comfort of a vehicle that can exploit the public roads. We are however by the same token, car dependent and addicted to what it is able to offer despite its real costs, restrictions and inconveniences. Cars are appropriated as a 'magical object' (Barthes 1957), fetishised and offer both visual and visceral pleasures combining 'mechanical impersonality with everyday intimacy' (Pickett 1998, 23). They are a form of adornment that attributes status and their 'sculpted appearances' are seen as objectifying sensation and desire. The 'elemental experience of speed puts cars into a different category from other commodities' (Pickett 1998, 23). Our relationship to cars and the sculpting of cars to suit desires also sets them apart from other objects of consumption. While they meet a fundamental need – that of mobility – cars are purported to offer much more:

> Australians and New Zealanders eat, sleep, fish, hunt, cosh and make love in motor-cars. They have babies in them, carry everything from live crocodiles to budgerigars in them. They have fought wars from some fairly rickety models, held concerts in them, committed murder in them, driven around seven million acre properties in them, and explored and developed both countries in them. They race each other up hills in them, around racetracks, down stairs and through places like Lake Mirrapongapongunna. And, of course, they sometimes round up sheep in cars. (Pollard, 1974, ix)

We ostensibly owe our lives, our present and the development of our countries to cars, and we are wedded to them. The range of activities they have accommodated have exceeded anything intended or dreamed of by the manufacturers and yet they are also limited and limiting. We have a very intimate relationship with cars that nevertheless allows us to overlook the shortcomings and inconveniences.

Cars are generally considered in quite separate terms from drivers in research relating to road safety. The focus is either on the psychology of individual drivers or the technology of the car and features that can increase safety. There is very little overlap between the two, and the symbolism of the car and how it is framed and articulated is given no significance in the sciences. In sociology there has been questioning of the neutrality of the car (Urry 2000) and some theorisation of the relationship between drivers and cars (Michael 1998). This relationship has seen the car considered as an extension of the human body (McLuhan 1964) or a more or less private space that is inhabited, but also as an intertwining of human and non-human, a 'hybridisation' (Urry 2000, Michael 2001, Hagman 2006).

The possible effects of vehicle characteristics on driver behaviour has been investigated in studies such as Horswill and Coster (2002) who found evidence for a relationship between vehicle characteristics and driver risk-taking using an observation based survey and questionnaire study. Despite more safety features in newer cars, crash rates have remained steady due to the riskier behaviour of drivers possibly compensating for the extra safety margins provided by the car (Fosser and Christiansen 1998). It has been noted also that there are different crash rates associated with different types of vehicle (Horswill and Coster 2002, 86). Horswill and Coster's (2002) first study showed an association between higher vehicle performance and higher choice of speed that was not simply due to the capabilities of the vehicles observed. Rather, 'drivers of higher performance vehicles chose to drive faster, even on a residential road with a low speed limit' (89).

In order to distinguish the causal direction of the relationship, that is, whether individuals with a propensity for higher risk-taking are more likely to choose a higher performance car or whether the design of the car influences the risk-taking propensity of drivers, another study was carried out. Horswill and Coster suggested that this was important because it determined who was obliged to take responsibility for risk-taking behaviour – manufacturers would have to take into account design aspects that could be shown to influence risk-taking behaviour. Drivers were asked about their risk-taking intentions in different kinds of cars. If a higher performance vehicle resulted in drivers being more inclined to take risks than other kinds of vehicles then it would show that the vehicle type was a factor influencing risk-taking behaviour.

The results showed a bidirectional relationship, that is, those who drove faster tended to buy higher performance cars and also that higher performance vehicles caused drivers to travel faster. A further study reinforced the relationship between vehicle performance and risk-taking propensities. It showed that increased performance and safety features correlated with increased speed choice, reduced car following distances and length of gap in traffic people were prepared to pull into (Horswill and Coster 2002, 101–102).

A study of car and pickup truck drivers found that pickup truck drivers, who were primarily males aged 30–39 years and married, had a lower reported restraint use and reported more risky driving behaviour and also had more traffic citations (Anderson, Winn and Agran 1999). The authors maintained that better safety messages needed to be targeted for pickup truck drivers.

Another study of vehicle types and driver risk-taking found that the 'higher aggressivity' of truck based Sports Utility Vehicles[1] (SUVs) and pickups makes their combined risk higher than that of almost all other cars with the exception of sports cars (Wenzel and Ross 2005). The study was focused on vehicle design characteristics and how the technology of cars could be improved. It also took into account the possibility of driver characteristics playing a part. Wenzel and Ross analysed fatality data in order to determine the risk-to-drivers and the risk-to-others of different types of vehicles. Next to sports cars, truck based (as opposed to unibody) SUVs and pickups had a much worse combined risk than other vehicles, which the authors largely attributed to vehicle design. The crash data was cross referenced with three particular driver characteristics with strong risk associations; driver age/sex, an illegal driving measure ('bad driver rating') and urban/rural driving (Wenzel and Ross 2005, 490). With some vehicles it was evident that age and sex could only explain part of the differences in risk-to-drivers but when combined with illegal driving, correlations between young male drivers and vehicle types were evident. Sports cars had the highest risk factor, the highest number of young male fatalities and highest bad driver rating. Pickup drivers appeared only slightly worse than the average car driver in this analysis – much worse than minivan and large car drivers and much better than sports car drivers (Wenzel and Ross 2005, 491).

---

1   Sports Utility Vehicles are compact four wheel drive vehicles based on rough terrain vehicle used in the country such as the Toyota Prado which is based on the Toyota Landcruiser but more often serving as family vehicles in the city.

Wenzel and Ross (2005) considered factors such as the likelihood that minivans were driven as family vehicles carrying children and would therefore be driven with more care, and that safety related marketing of European cars, attracted safer drivers. SUVs on the other hand, it was noted, had not been sold as requiring the special care in driving required for trucks. The authors considered these important factors to take into account in considering the influence of driver behaviour on car risk rating based on crash data. The higher risk rating of pickup trucks was subsequently related primarily to design and being driven in rural areas and less to driver behaviour (491–492).

One other study that is of interest here because of the way the car/driver relationship was approached, is the Benfield, Szlemko and Bell (2007) study of driver and vehicle personality attributions. The study considered whether a predictive relationship could be established between characteristics attributed to the car and aggressive driving tendencies. They were surprised to find that of the 200 undergraduates who filled out the personality inventories for both themselves and their vehicles, none questioned the attribution of personality characteristics to a car.

The analysis showed that driver and vehicle personalities were distinct and not just the projection of the driver's own personality into the vehicle. In the correlations of driver and vehicle personality with indexes of aggressive driving tendencies, some were predicted better by vehicle personality than driver personality. High extraversion, low agreeableness, low conscientiousness and lower openness were associated with more aggressive driving, and vehicle agreeableness was negatively correlated with driver anger (254).

These studies all represent attempts to consider and separate the car/driver relationship and yet none of them take into account the symbolic dimensions of the car or the connection between car and driver in any substantial way. The last study (Benfield, Szlemko and Bell 2007) allowed for a relationship between car and driver more seriously but this could only be considered as anthropormorphism, that is, attributing human qualities to an inanimate object. Cars are more than inanimate objects just as they are much more than a means of transport, and the importance of the study is that it shows that the qualities attributed to cars exceeded the driver's own personality. The relationship between cars and drivers is more complex than that between an inanimate object and a person and cars have a 'being' in an important sense.

The car is of significance independent of individual users and provides a real physical presence but there is also an important interconnection between cars and drivers. Tim Dant explored the assemblage of driver-car as a form of social being that produces a range of social actions through the idea of 'affordance' recognising that the car shapes and is shaped by human action (2004, 61). Dant flags the difficulty of theorising the relationship between human and machine in his discussion of the notion of affordance and alternatives that draw on notions of hybridity and cyborgs. As an affordance a car gives mobility (movement) and motility (spontaneity and independence) but it also 'enables a range of humanly embodied actions available only to the driver-car' (74) Dant's analysis draws on Merleau-Ponty's phenomenology to allow a more effective encapsulation of the car-driver assemblage as an embodied experience whereby humans have adapted to the form of perceiving brought about

Figure 1.2   Fear No1: a customised car type popular with young men affords an identity

through the car. The car is a shaping force though lacking in intentionality but it is the driver-car that as an 'assembled social being' with properties of both person and thing that is meaningless without both (74). This social being will be explored further through the idea of articulation to be outlined in the following however it is not so much specific perceptions brought about through the car that will be of interest here but the meanings applied to cars through its assemblage with particular humans. First I will give a brief history of the social significance of the car and then I will outline the importance of social framing through the media and gendered relations to technology.

A short social history of the car

The changes in culture that were wrought by and with the car are significant in establishing the enormous impact of the car and its connection to changes in the way lives were lived. While the car developed with other technologies of the home, industry and consumption – radio and television reaching into people's homes, separation of work and home, and the growth of shopping centres – it did not merely permit or facilitate those developments but was in part responsible for them (Urry 2000, 59). In the social patterns of life that developed automobility was a key feature. 'The car's significance is that it reconfigures civil society involving distinct ways of dwelling, travelling and socialising in, and through, an automobilised time-space.' (59)

 Cars were originally a toy of the wealthy who enjoyed the speed and status it offered. An obsession with speed developed with the car as it was able to propel the rich ever faster. There was a desire to set land speed records, even though there was considerable controversy over the costs and benefits (Urry 2000, 60). For the ordinary worker cars were financially out of reach and embodied the arrogance and self-centeredness of the wealthy:

> Cars met with particular hostility in densely settled cities, where people living in crowded tenements used the streets for games, socialising, buying and selling, and other activities that had nothing to do with transportation. (Volti 2004, 18)

They also met with hostility in the countryside where people complained about 'reckless drivers, the dust kicked up by cars, and the proclivity of motorists to trespass on private property, steal fruit and strew trash.' (Volti 2004, 18) Extreme reactions involved digging up roads and putting barbed wire across highways. With the expansion of car ownership however, hostility began to subside.

Once motor vehicles were mass produced with prices the middle classes could afford they were embraced with enthusiasm. This lead to a new phase in which the emphasis was on finding slower means of enjoying the pleasures of motor touring: 'To tour, to stop to drive slowly, to take the longer route, to emphasise process rather than destination, all became part of the performed art of motor touring as ownership of cars became more widespread.' (Urry 2000, 61) The car became an extremely desired item in the late 1940s and by 1965 an automobile culture had taken hold in the industrialised countries (Volti 2004, 88)

Little attention was paid to safety, braking power and suspension in 1940s and 50s with greater interest in style and colour and a 'longer-lower-wider' theme (Volti 2004, 97). The dominant styling idiom borrowed heavily from military jet aircraft – 'wrap around windshields and roofs shaped to look like cockpit canopies, fake air scoops, and exhaust pipes made to look like the tail end of a jet' (98). This was combined with heavy doses of chrome weighing the cars down. At this time American auto design was so concerned with style it had become almost completely detached from the primary function of transporting people safely and efficiently because this was attractive to the buying public and manufacturers (98). The excesses of design reached a peak in the late 50s however and buyers turned to smaller more restrained designs including foreign cars, particularly the Volkswagon (99).

An image of high speed car travel had been planted when in 1939 General Motors' exhibit at the World Trade Fair showed an image of the American landscape in 1960 featuring a grid of 12-lane highways where cars travelled at high speed with no traffic lights or congestion to slow their passage (Volti 2004, 106). Plans began after the war for a new highway system that might turn this vision into reality. Widespread support for the development of a highway system reflected the unquestioned acceptance of automobility. Nevertheless highways with limited access devastated urban areas and divided neighbourhoods as homes and shops were destroyed to carve a path through the middle of suburbs for the new roads, though this was of little concern to those who glided by on the highways in the comfort of their cars (107).

From early in the development of cars there was a perceived right on the part of those who owned them to take priority on the roads and to travel at the speed they saw fit. Responsibility for crashes on bends and curves was from the earliest stages,

Figure 1.3 **Preparing for higher speeds. A new freeway carved through the bushland north of Sydney in the 1960s now carries thousands of commuters daily between Sydney and the central coast. Photo by Ken Redshaw**

blamed on the road builders and government planners. In 1925 in Australia it was noted by the Assistant Chief Engineer in New South Wales that the requirements of road users had changed, primarily in demanding a smooth, weather proof surface that would increase the longevity of their costly motor vehicle and the comfort of those travelling within it. Drivers also demanded 'the absence of sharp or screened curves, with their hidden perils, and their need for slackening speed' (Broomham 2001, 112). There was consensus on the part of authorities that the roads should meet the demands of the driving public. The Main Roads Board in NSW stipulated that roads be made as straight as possible with wide curves and moderate grades to avoid the crashes that occurred on sharp curves.

Safe roads also came to require increasing numbers of lanes. Roads with enough space for vehicles to pass each other were further developed with the bitumen sealing of roads to weatherproof them. As motor vehicle numbers increased it was necessary to add overtaking lanes so as to circumvent the impatience of drivers caught behind slower vehicles who were inclined to take risks to get past. An illustration of a climbing lane on the Pacific Highway in New South Wales is accompanied by the caption: 'This climbing lane on Pacific Highway south of Gosford was the first of many. They allowed a steadier flow of traffic on highways where slow and heavy freight carriers caused queues which too often led to dangerous driving.' (Broomham 2001, 145).

As roads have improved, motor vehicles have travelled faster requiring further improvements and measures to lower crash rates. Even in the early days claims were made about the ability of cars to go anywhere. In Sydney competition for sales led to stunts in the mid 1920s such as showing an Overland 'jumping' a fallen bridge and another of a Willys Knight 'climbing' the steps of Sydney Town Hall (Pollard 1974, plates by Don Harkness shown opposite page 81). The reality was a very different one of dusty and muddy streets and roads, and the vulnerability of cars to weather and geography. Tyre blowouts, broken axles and overheating engines were common occurrences. In principle cars could 'go anywhere' but in reality they require expensive grading and sealing of the landscape to allow them to access all areas.

The car speeding along the open road has become a metaphor for progress and the taming of the wilderness (as documented by Wilson in the United States (1992). Surprisingly little has been written about the conquering of nature in Australia though motor vehicles played a key role, not just in the sense of making more areas physically accessible but also in promoting the car as pitched against the foibles of nature. Cars appeared to make great strides in taming wild unpredictable nature while they unleashed a storm of destruction of natural habitats and landscapes. Nature was seen as the problem, not the car, as roads paved the way through all kinds of terrain to make it more penetrable for motor vehicles. Nature was forced to submit to the advance of the motor vehicle.

The promise of the automobile has been, and continues to be, based more in an ideal world than the real one:

> In the ideal world the car will let you go wherever you want, whenever you want, at full speed, on an open road. There are no limits to our freedom of movement, no speed limits and no restrictions ... The image of the car mediates the car as a modern 'time-space-machine', compressing time and shrinking distances. It promises to take us further away in less time, in any direction at any time, and with complete self-control and independence. The 'automobile' promises to make the driver both 'auto' and 'mobile', that is, to give its user both autonomy and unrestricted ability to move around. (Hagman 2006, 66)

The automobile is an ideal that does not exist and the Ottomobile, named after the inventor of the internal combustion engine, delivers congestion and must be serviced by an endless string of 'servants' from parking officers, to car builders and salesmen, gas station attendants, traffic policemen and even medical services (Hagman 2006, 72). Hagman points out that advertisers do not show the obstacles, restrictions and obligations or the traffic jams encountered in traveling to work facing motor vehicle drivers. Rather, they show the open road with no other vehicles, the fun and excitement of a party, a holiday or other leisure activity (66).

The internal combustion engine or Otto engine is one of the foremost limiting factors making cars heavy and more destructive on impact. Urry argues that cars in the future will be made of lighter materials because they will not have to carry around the enormous weight of the internal combustion engine which in turn is required to propel the weighty metal vehicle (Urry 2004). This highlights yet another ambiguous feature of the car since the presence of heavy metal vehicles means that lighter vehicles might be more vulnerable if hit by today's vehicles making a transition to a lighter vehicle more difficult. Transitions to alternative vehicles will need to be carried out in city streets where moves have already been made to restrict the Ottomobile to overcome congestion and to give some freedom back to pedestrians.

Another feature of the car that requires mention here has to do with the kinds of privacy and individuality fostered by the car. The privacy of the vehicle is tempered by its presence on a public road where others are able to see in, and the individuality delivered by the car tends to be one that pits one competitively and aggressively against another and depends on a market where goods are mass produced. In fact, the car has been in part responsible for forcing the extent of individuality and privacy upon us (Urry 2006). These values are as unsustainable as the internal combustion engine in the future. Wealthier countries and individuals have adopted wholeheartedly the convenience of a vehicle that is available solely to ourselves to dispose and use as we wish but this is a luxury that we cannot continue to support.

Many of us would happily opt for a form of mobility that did not require finding temporary spaces to accommodate it everywhere we went. If for example we could access a vehicle that would take us where we wanted to go in relative privacy and then disappear when we arrived at our destination and return when we wanted to leave, we would happily adopt such a form of mobility. Cars are so implicated in our notions of freedom and individuality that these issues will have to be addressed and our ideas of freedom questioned and unpacked in order to bring about changes in our very 'private' yet nevertheless public experiences of mobility.

Social framing

Before going on to show how cars are made meaningful through articulations that urge particular behaviours, it is important to consider the way the relationship between media and objects of consumption have been theorised. In psychological research attempts have been made to prove a causal role between car advertising and particular driving behaviours, however, such a causal connection is notoriously difficult to prove because the relationship is more complex and less directly a result of any one particular advertisement.

In media theory other approaches have been taken to the relationship between media and society. While the overriding interest is on the impact of media power in society more broadly, the issue has been approached in different ways. The examination and interrogation of media texts and institutions has been important for the theorisation of media impact but limited in what they can say about social consequences. Media theorists then turned to the question of what people make of media products and how they are used in social life. Even in Marxist approaches examining the structures of media production and concentration, there is a crucial uncertainty as to how media texts 'causally mediate between the world they represent and the world where they are consumed' (Couldry 2004, 118). Audience research attempted to investigate the question of people's consumption of media but this was constrained by a fundamental emphasis on people's relationship to media texts. Silverstone (1994) argued that television has been integrated very quickly into the routines of everyday life and it is the familiarity of these routines that are the source of our sense of security (Carrabine and Longhurst 2002, 186). Different questions were posed in media studies such as 'what it means, or what it is like to live in a media saturated world' (Ang 1996, 72). Couldry has also given emphasis to the size of media institutions thus including power in analyses of media (Couldry 2000). Couldry's proposal is to start with 'media-oriented practice' in order to consider more broadly what people are '*doing* in relation to media in a whole range of situations and contexts' (Couldry 2004, 119) and not just what media texts are 'saying' and how they are being 'read'.

Given that most people are receiving their news, a great deal of information and knowledge about consumer items, and dramatisations of life through the media, mass media can at least be considered to be involved in social framing, that is in shaping how and what consumer items are being used and made meaningful.

'Frame' refers both to 'the devices through which media texts structure or 'frame' our perception of the realities they represent' and 'the framework of expectations and plans which are associated with, or 'frame', an action'. (Couldry 2000, 16) Social framing then, in which the media plays a significant part, generates, reinforces and determines expectations and beliefs which limit the options available to those that dominant groups such as manufacturers, governments and large groups in the population prefer.

Other aspects of social framing include social norms such as gender. Men have been dominant in the generation, production and use of technology and have in turn generated technology, which suits male ideas of power, use and form, and these ideas have in turn been 'baked' into the technology. Women are expected and believed to be less competent with technology than men:

> The 'woman driver' cliché, so often ringing in our ears, helps to produce an unconfident woman driver and contributes to the making of men in the persona of 'driver', a driving person. (Cockburn and Ormrod 1993, 1)

If car shows where manufacturers show off their newest designs are any indication, cars are determined by male obsessions with speed and power. Where cars today have sufficient power for most purposes and speeds in cars have reached an optimum level in the balance between costs and gains in time, manufacturers are consistently advertising more power and acceleration, higher top speeds and performance. Women meanwhile are more likely to be interested in convenience and the spatial use of the inside of the car such as where drinks can be easily placed and other items stored for ease of access. This aspect is rarely a strong advertising point for cars even though manufacturers have increasingly addressed storage spaces in cars more effectively, recognising women as consumers. Men pride themselves on their ability to control powerful cars and while driving powerful cars and driving them well are considered important personal attributes, especially for men, considering control of a washing machine as skillful in the same way would be laughable (Cockburn and Ormrod 1995, 1). This also swings on the potential of the car to kill and the danger that it poses which is both valued and denied in the promotion and masculinity of cars.

These are all important aspects of social framing. While cars are technologically regarded as neutral they are nevertheless socially framed as essentially masculine and the domain of men, as is much technology. Cockburn and Ormrod's (1993) study of the microwave oven from design, manufacturing and marketing through to use in the home, showed the inscription of gender in every level of the technology and the value inscribed in masculinity through the technology. The relative importance of men and masculinity evident in their study 'was not an artifact of the microwave-world alone', which may have embellished and enhanced it, but of wider social patterning (159). They point out that competent and frequent use of technologies, as well as intimidation by technologies, shape identities. Men and women differ in the kinds of technologies they are comfortable with and intimidated by such that technology is gendered and in turn genders (159). Gender will be discussed in more detail in Chapter 3. Here we only need to note that gender is a significant aspect of the social framing of the car and the meanings that are baked into it. Cars are indeed gendered in significant ways in their very design.

Articulations of the car

Articulation applies to cars in the sense that cars are seen as expressive and used in expressive ways to convey particular meanings. It is not just the way cars are spoken about but also the meanings that the appearance or 'look' of the car is intended to convey through the way it is built as well as how it is framed and promoted. It is not just the expression that is built into cars that is important but also the social framing of cars by manufacturers and in the media. Cars consist of connected parts that make up a coherent whole, capable of action as well as 'saying something'. Much the same as body language, cars are articulated in particular ways to express and make possible different actions and emotions. Male bodies are articulated in ways that convey what is culturally considered 'maleness' such as sitting with legs apart, an articulation that female bodies are not expected to express. Similarly cars can convey 'friendliness' or 'meanness' and aggression, and cars and colours can be considered male and female.

An early example of how meanings can be structured or 'baked' into cars is the controversy over the inclusion of an automatic ignition or self-starter instead of the dangerous crank handle on cars. It was claimed that men would find the ignition too feminine and so it was associated with a concession to feminine convenience (Scharff 1999, 60). Early electric cars had already been associated with women as had additional internal comforts such as softer suspensions and more comfortable stylish leather seating while internal combustion engine powered vehicles were associated with men (55). Though men liked to boast about their victories over early cars which were often difficult to handle, Scharff says 'it is by no means clear that ordinary male motorists disdained comfort, or that they always enjoyed demonstrating masculine prowess by deliberately choosing a Spartan automobile' (55). In spite of its instant and universal acceptance the self-starter was nevertheless promoted as a feminine convenience enabling women to take the wheel (63).

Figure 1.4   Citroen Classic a car of multiple meanings. This model was known for its sleek design suggesting speed

A variety of expressions are exemplified in people's experiences of cars and styles of driving often connected with particular cars. Enthusiasts[2] in their latest Falcon XR6s and 8s or Holden SS or SSS Commodores could be considered car centred as they speed along in the overtaking lanes and dodge into other lanes as their progress is slowed and the commuter in an ordinary sedan with no trim, the family wagon or people mover sticking to the middle lane could be considered mobility centred. Tourism involves using the car as a means to enter and explore, for enthusiasts it involves the experience of the vehicle on the road, pushing the limits and penetrating the harsh environment. The commuter values convenience and comfort and perhaps privacy in maintaining the use of the car. These various experiences merge somewhat in the flow of traffic where the central issue becomes moving ahead.

Often driving is associated with thrill and excitement, especially when it comes to flash cars and speed. Enjoyment in these circumstances is associated with the idea

2   'Enthusiast' is being used too broadly here but it is intended to cover those who are more car identified and it is recognised that this can include those who simply love particular kinds of old cars as well as those who like to demonstrate their driving prowess on the roads everyday. Hoons or boy racers are usually associated with younger men but can also be older and driving like a hoon in the latest hotted up version of the Ford Falcon RX6 or 8 or Holden Commodore SS or SSS. I have stuck to using enthusiasts to include them rather than trying to stretch hoon to a different age group. Apologies to the innocent lovers of cars of all types who do not need to show off by driving aggressively on the roads.

of free flowing movement, unobstructed by traffic, where speed is open. Responses to the restriction of speed and the impediments of traffic can be aggressive, and fast cars and speed are often associated with aggression. Pleasure is derived from passing all the other cars along the way and at speed additional to what is prescribed.

The association of cars with mobility in the sense of domination is not far below the surface of many representations of cars. The commuter can come to feel they are being left out or left behind in the traffic as others seem to get ahead in the flow of traffic, pushing their way aggressively forward. While the car in itself dominates the environment, aggressively carving its way through all social and cultural contexts, dominating by its sheer weight and force, it can also be made to look and 'act' more aggressively. Cars can be driven to express wild and challenging styles of driving that are aggressive and this can be the aim of car advertising in its contribution to the articulations of cars.

There is a lot of aggressivity associated with race and rally cars which are designed for highly competitive driving and are intended to look aggressively competitive and exciting at the same time. The magazine advertisement from *Wheels Magazine* shown in Figure 1.5 (August 2003, inside front cover) shows the predominantly red car as two sided – the same car is both a rally car and a road car. The meanings associated with the rally car are being transferred to the road car. The drivers dressed in the full red and white rally suit add to the racy appeal. A rally car is driven at speed with a measure of reckless abandon, sliding and drifting around corners in a controlled but hazardous fashion. Rally drivers, as with race drivers, are considered extremely skillful drivers who live on the edge and experience the real meaning and virtue of cars – to race hard and fast. Race driving is about the driver and the car and beating out the competition. As such it is aggressive and does not involve care and consideration of bystanders who must be kept separate from the vehicles hurtling past.

The text from this advertisement for Toyota's Corolla Sportivo states it's a 'true driver's car', and uses terms such as 'awesome' and 'thoroughbred' relating it to both rally driving and circuit racing:

> The rally-inspired Toyota Corolla Sportivo. With an awesome 1.8 litre WTL-I engine, six-speed gearbox, 16 inch alloy wheels, leather seat trim and fully colour-coded body work, it's a true driver's car. And the next generation in a long line of thoroughbred Toyota racing cars.

Another advertisement for the Sportivo emphasises the 'thrill', unbelievable power and racy technology and looks:

> The adrenaline-fuelled Toyota Corolla Sportivo. With an aerodynamically efficient body and the unbelievably powerful ZZZ-GE engine, it's every inch a driver's car. This 1.8 litlre WTL-I engine (variable valve timing lift with intelligence) delivers a massive 140 kw of power through a six-speed close-ratio gearbox, and the Sportivo is the only Corolla that has it. Sports suspension and brakes, fully colour-coded body kit, sports alloy wheels, sports front seats and leather seat trim all come as standard. From just \$29,990, nothing will ever be as thrilling. (Wheels Magazine, August 2003, 46–47)

Figure 1.5   Toyota Corolla with rally associations

The performance and power of many cars are emphasised as 'rally bread' or as 'race bread' with car companies scurrying for the upper hand in technology on many levels, from the engine capacity and handling on corners, to the inside comfort and safety. The features of the car take on an aggressiveness because of their association with the driving styles of racing and rally driving and yet these styles are completely inappropriate as models for road driving (Redshaw 2007). Toyota intends the Corolla to appeal to both men and women so it has a 'friendlier' grill and lights on the front but the red colour and the association with the fun and the challenge of rally driving maintain a masculine appeal.

Now that the fashion for low petrol consumption has subsided, the emphasis is even more on the power of the V8, which is being produced again by Ford and Holden and also by makers such as Saab and now Toyota. Honda flashes a race car across the screen at the end of many of their ads. Mitsubishi shows both its Lancer and Magna on the rally circuit in television advertisements. Somehow the violence of the race and the rally, and their inappropriateness as a model of ordinary driving is overlooked. The race exemplifies the challenge and the potential for loss of control that appeal to some men as drivers. This is a significant part of the articulation of cars as masculine throughout the history of the car and how it has been promoted.

The ways in which the car is articulated allow and tolerate the violence that is incurred in order to maintain individualism in its dominant expressions – as competitive, dominating and aggressive. The dilemma is the increasing need for control to manage the spiralling force and presence of cars. The car is promoted as the means and emblem of individual expression and performance, power and speed are standard features of many cars. More power, better performance, faster acceleration are constantly being generated and promoted to make the car and its driver stand out. Sports packs on cars are increasingly popular making them look more like 'real' race cars. The race car has moved further and further away from the road car and yet the alliances are being emphasised more than ever. Formula One cars are a breed of their own, but they are related to the road car through technology.

Another campaign that is worthy of note is a Ford billboard campaign displayed on a motorway in Sydney for Ford Performance Vehicles in 2005. The ad was placed on a bridge over the motorway so that the viewer has to look up at it as they pass below. The viewer is placed in a position of vulnerability in relation to the vehicles with the front end of the cars seen from the perspective of someone about to be driven over. The implication is that if you are not in the car you might suffer the consequences and only get to see it from the angle of being run over. The front of the car is intended to suggest aggression and further adding to the aggressiveness of the ad is the number plate on one of the cars – Typhoon. Performance in the ad is clearly related to aggressiveness. The grill could be considered a snarling set of teeth and the body mouldings around the wheels suggest well muscled power and grunt. The streets are framed as concrete jungles in some adverts with the implication of unruliness and 'everyman for himself'. Others such as the high class Holden Statesmen are marketed in adverts that suggest the roads are combat zones and that the car is able to dodge whatever is thrown at it and protect its occupants in comfort and security.

The car and the aggressive uses of the car are taken very lightly and its effects on others and the social environment underplayed in the articulations applied to cars such as those evident in advertising. The priority given to cars over many other aspects of the social environment is demonstrated in the aggressive pushing through that is implicated in much current car advertising. The advertisements discussed here serve to illustrate one of the problematic themes of driving discourse – the unquestioned priority of the car and the lack of recognition of its destructive potential.

The struggle to articulate and define the legitimacy of coercion on the roads appears to be related simply to what the car can do and not countered by the destructive consequences that can follow from the car. The ability of the car to go faster appears to overrule the cost and appropriateness of faster speed. It is then addressed as a tool for saving time at all costs, and for getting ahead and the extent of the symbolic significance is hidden.

The masculinity of large performance cars is often emphasised in car advertising. A number of adverts in Australia have shown men driving performance cars such as Holden's SS Commodore and Ford's XR Falcons around challenging bends on isolated open roads. One advert shown in 2003 showed a young boy sliding down the curved banister of a stairway and then a youngish man driving the car around sharp bends with the words from a song 'I don't wanna grow up ...' (Holden Commodore). This conveys the idea that men naturally like to experience the thrill of a powerful car but also that the car itself is masculine enough to meet the desires of men. There is no concern for the effect of speed and the experience of the self centred thrill on the surrounding environment. The impact of the speed and power of these vehicles in traffic contexts and on suburban streets is denied in the emphasis on masculine power and the fantasy of the open road.

Figure 1.6   Yellow Ute. A vehicle particular young men prefer

Television adverts for utilities used by tradesmen in Australia but also recently popular amongst non-trades men because of its identification as a 'man's' car, have shown the car as muscly and powerful enough to take the knocks of hard work and cope with various kinds of terrain. A current utility vehicle advert shows a man hooking a tree stump up to the ute and then revving up just enough power to pull it out. An inserted scene shows a tree disappearing from a square in China or Russia, which then appears in the back of the ute. The man's expression indicates his manliness in being unphased by the appearance of a tree attached to what had only appeared as a stump. He simply gets in the car and drives off with apparent disregard for any impact the 'awesome' power of his ute might have had on the other side of the world.

An article in local newspapers about the Dodge Nitro describes it as a 'quality vehicle with killer looks' evident in its aggressive styling, alloy wheels and performance suspension. The Nitro will allegedly 'deliver "that little something midsize 4WD buyers are still searching for – attitude"' (Crawford 2007). The Dodge Demon roadster meanwhile is described as muscular with features such as its '19 x 8-inch alloy wheels and its frontal styling and headlight treatment' which 'combine with the signature Dodge crosshair grille to give the car an aggressive persona' (Crawford 2007).

The Mercedes C-Class on the other hand is described by the design team as having taillights shaped to make the car seem wider than it is, giving the impression that it is 'elbowing someone out of the way'. Mercedes are attempting to appeal to a younger buyer and apparently these 'young fast-paced sharpies … want a car that can dart in and out of traffic. Previous C-Class sedans have had a more sedate feel. The new one is tuned for faster steering responses and more agile dynamics …'.

The articulations of cars build social values into the very body and mechanisms of the car highlighting the central importance of gender, class and social standing but also more specific characteristics of the driver. Some men do not want to be associated with a ute which is more the choice of trades professionals and those who want to be associated with physical ability and competence. Lawyers, doctors and other professionals are more likely to drive sedans and sportier vehicles. The car is

Figure 1.7   Dodge Nitro grill

considered to say something about the driver to the extent that for some it would be an embarrassment to be seen in another type of car.

Cars as incentive: suggestive cars and driving styles

Cars then are claimed to suggest aggressivity, to possess and convey 'attitude'. In the Transforming Drivers gender study (Redshaw and Noble 2006) focus group participants were asked whether they associated particular kinds of people with particular kinds of cars and whether they drove differently in different cars. While many denied that the type of car you drove would influence who you were, as one woman stated, 'it can influence the way you drive, but not who you are' (Female, Goulburn), many of the young people came up with examples where the type of car effected how they drove. It became apparent that particular cars require being driven in particular ways, or coerce the driver to conform to the specific meaning associated with the car. Acceleration or take-off power is often emphasised and is important in the articulation of larger cars. One young woman noted how she felt driving her father's car:

> If I get in my dad's car, which is a Commodore, because it's an automatic, I do drive differently but I don't drive recklessly. It does have the power to go faster, so you tend to take advantage of that. (Female 3, Blacktown)

Although she realised she did take advantage of the ability to go faster and to drive differently, this woman said in her defense she 'does not drive recklessly'. She only drove the car according to how it could, and some would say, is meant to be driven. Similarly some of the men said that being in a nice car would make them feel 'more confident' and want to 'try something' and give them 'incentive to go faster' (Goulburn). Even though they state that the car encourages them to drive in a different way, they deny that it is influencing them.

Young women were often more reflective about car types and their associated meanings than the young men. Many young women referred, sometimes in derogatory terms, to the kinds of cars they associated with young men. One woman referred to 'young guy cars which are the 'look at me' type things' stating that 'the sort of car you have can reflect what sort of driver you are, especially with young drivers'. She said you could sometimes tell if a young man or a young woman is driving a car by what car it is, 'and you expect to know what sort of driver they are' (Female 1, Blacktown). While the women were more reflective, they did not necessarily make an explicit reference to gender. A woman in the Bankstown group referred to one of the male participant's cars as 'a popular, young car. It's one of those done-up cars. It's a fully sic car! (laughter) With a new paint job!' This was not the kind of car many women were likely to profess a desire for.

Sports cars 'require' being driven in particular ways that show off their handling and performance, and that of the driver. As one young woman put it, 'if you're a really sensible driver, you're not going to go and buy a fully sports car that's going to go out really fast' (Female 1, Blacktown). Young men it was clear, were associated with and liked to associate themselves with, 'hot', flash and powerful cars that often involve 'modifications' to the original design, and were either big cars such as Commodores or sporty cars with plenty of power. A young man from Wollongong stated: 'Even if you can't afford it, you'd still choose a nice car. Something sporty, because it says something'.

Young women were happier with smaller cars that used less fuel and they did not see the need for fast cars or high performance cars when 'you could only do the speed limit anyway' (Goulburn). They expressed particular, usually more cautious driving styles. They did feel nevertheless that cars were expressive for them as well:

> Yeah, it's more sort of what expresses you. Like, I think my car suits me pretty well because it's sort of cute and run-around and stuff like that. And I've got friends who are big, butch guys in tiny little bubble cars and it just doesn't work. (Female 1, Lawson)

One woman said she liked 'little cars, they're fun' but went on to say 'except everyone thinks that they can mess with you, and then they find out they can't!' (Female 3, Lawson, 17 years) Her driving style, she suggested, made up for the way she felt she was treated driving a small car. A young man from Wagga felt he could make up for the fact that he was driving his girlfriend's small car, which had a pink interior, by 'hooning around in it'. He said he would only drive it if she was with him. Some car types it was felt then, required being driven in ways that were contrary to expectations expressing qualities of the driver. Many of the men pointed out that even though they drove small cars, 'they've got power, you've just got to know how to use it' (Goulburn):

> Yeah. I know a really good friend of mine, he bought an Excel, a brand new Excel, and he's probably six foot two, and plays league, and they were all like, 'You bought a Barbie car!' (laughs) So I think smaller cars you sort of associate with girls. And if you see a girl driving a family car you think they're married, or they're about to have a baby! I'll stick with my little car, thanks! (laughs) (Female 3, Goulburn 2)

For young women the car could give the wrong kind of message about them so they would be more inclined to drive a single person's car than one that suggested they had a family. Small cars like Holden Astras and Barinas, Hyundai Excels and even small SUVs were popular amongst the young women.

Cars could be defined as male or female depending on a variety of factors including size, style and performance, but also whether it had been modified and how you drove it. Young men in their initial enthusiasm for cars put a lot more effort into them than when they are older, spending Friday and Saturday nights taking their cars to popular places to show them off by cruising around a block, displaying the modified body work and possibly their driving skill, stopping and lifting the bonnet to discuss the modifications to the engine with other young men, and playing their car stereos as loud as possible. One young woman described her boyfriend's current interest in cars, which included displaying his car at The Rocks in Sydney, a popular hang out for young men and their cars:

> My boyfriend has a hotted-up car as well, so we go cruising. He's down at The Rocks every Friday and Saturday night showing off his car. If not, he'll go to Wollongong, Bondi, Brighton. All the young spots for the young, because he's a young guy. All the hot spots where all the fully sic guys are! That's what his weekend consists of, and I hope he snaps out of it soon! (laughter) (Female 3, Bankstown)

It is clear that gender and age are playing a part in the performance of driving recounted here, as it is more commonly a male preoccupation and many of the participants are under 20 years of age. The type of car is important, as there are particular specifications if participating in Friday and Saturday nights at the Rocks, and will involve particular styles of driving to go with them. 'Hotted up' Hondas, Subaru WRXs and Mitsubishi Lancers are popular for these young men in the western suburbs of Sydney.

Young men in an all male focus group discussed a television ad for the Holden Monaro, a high performance car that is shown as unapologetically competitive in a computer game format, racing traditional high performance sports cars, through city streets and along dramatic cliff tops. The game format is represented at the beginning of the ad when it appears that the viewer gets to choose the colour and other features of the car. The young men clearly saw the car as a male car to the point that it was even seen as necessary for guys to have that kind of power:

> M2: Well, that's the whole point of having a Monaro in the first place. It's a V8.

> M3 It's a guy thing, you need it, you know what I mean? It's for show.

M1:   It's more, like, a testosterone thing. So obviously they make it for a reason, so they have to show if off one way or another. (Fairfield)

The young men recognised that the car was made for males like them and that it had to be shown off in the way it was intended to be driven. Young people in other groups remarked that the ad was 'just promoting speed and power' and on how the Monaro had to be driven: 'I'd go fast if I had a Monaro' (Female, Bathurst 2). A male meanwhile made it clear what he thought the intended articulation of the car was:

> You don't buy a Monaro to drive around at sixty kms an hour in town. You buy it to drive fast. It's a fast car. It's powerful. (Bathurst 2)

Young people are certainly susceptible to the 'drivability' of particular cars and their ability to demonstrate it. If you have a flash car, you are going to drive it in ways that show it off since, as some of the young women put it, 'You don't want to kill the cred', and 'If you've got a fast car, you want to drive as fast as you can!' (Wollongong). In this sense the car itself, with the particular meanings that are 'baked into' it can be a coercive factor influencing how it is driven.

Car size

A characteristic of cars that was very important was the size of the car and how it related to the way the driver felt and this was explicitly discussed by the young people:

Figure 1.8   Subaru's WRX. Another car liked by young men for its speed, power and performance

In my mum's car, I feel more … I feel sort of more grown-up, because it's bigger, it's a family car, whereas even though obviously I haven't got a family, in the little car I feel more like a teenager, whereas with the big car it's more a family car and there's enough room to actually put four people in it, plus a baby … (Female 3, Lawson)

Car size could be an important factor in creating feelings of frustration with smaller cars considered slower and less powerful in general:

When I drive my mum's van, I feel a bit different. If I was driving a little car, I'd just get frustrated, I think. (Female 1, Lawson)

It was important that you could not expect as much as in a small car and then in a larger car you had more to deal with in terms of performance as a driver:

Yeah. It's like, '*#%@!, I've got a lot to live up to. I better go a bit faster!' (Female 3, Lawson)

The size of the car could also affect more explicitly the speed at which they were driven. Even though they most often said they drove more carefully in their parent's cars, parents often drove bigger, more powerful cars and this influenced how the young people drove:

Yeah. That's the difference. In my mum's car, I'm always going at least 10 (kms/h) over the speed limit. It could be that because it's an automatic, it feels slower. So I'm like, 'No, I'm not doing 50.' So I'll do, say, 60, whereas in the little car I'm always doing the speed limit. (Female 2, Lawson)

The size of car related quite explicitly to gender. Young women were quite aware of how the more powerful cars made them feel and hence drive differently. For young men it often seemed an obvious choice that men would drive bigger and more powerful cars. One young man recounted the reaction of his friends when he changed his choice of car, having originally owned a bigger car:

When I said I was buying a small car instead of the VP Commodore, everyone said it wouldn't suit me, that my car matched my personality! (Male 3, Shellharbour)

This young man maintained, as did others, that he had made the obvious choice for himself but that did not mean his identity was affected by his choice of car: 'It doesn't necessarily make me different, but you definitely get – well, in my case, I definitely have a car for me. I haven't gone out and got something that's … a pink beetle! (Laughter) I've gone out and got a chunky VP Commodore.' (Male 3, Shellharbour) This was the first car owned by the young man, a large, powerful, older car that he could afford. It was commented by another member of the group that the same young man now drives a Nissan Pulsar which is a smaller car, and he responded:

Yeah. And people are just blown away that I've gone out and got a little, small 1.8 litre … they don't get it! (laughs) It just doesn't fit. They say it's not right. (Male 3, Shellharbour)

The size of the car is clearly seen as saying something particular about the driver and can be compensated by the style of driving, though as the young men got older most were less concerned about the power or size of the car and more concerned about what it cost them to run. Still they continued to some extent to maintain the idea of the car as saying important things about themselves as males.

Cars behaving badly

The boundary between cars and the people who drive them is very hazy particularly when it comes to apportioning responsibility. Cars are involved in the identity and gender and other aspects of the driver. They thus have an important function in the shaping of identities and the social framing of gender as well as being inscribed with gender and other characteristics. The new Mercedes C-Class with its suggestion of 'elbowing others out of the way' draws on class boundaries and the idea that those with money are able to seize priority and push aside those less worthy.

In focus group discussion of a safety campaign ad to be discussed in more detail in Chapter 3, young men consistently identified with the idea that the car could make them go too fast. Young men also often reported the enticement of a straight stretch of road which 'made them' want to drive faster and often they could not see that driving faster just because the road was straight could be a problem. They included little consideration of the context of the straight stretch of road. At the same time the car as a force influencing them is openly denied since they see themselves as in control but we have also seen that advertising relies on the potential for the car to seize control from the less skilled driver.

As cars become more 'forgiving' with better steering and cornering ability, better braking and even responding to closeness to the vehicle in front, the skill of the driver is emphasised even more suggesting that the technology is taking nothing away from the skillful driver. Cars must present a challenge that the skilled (male) driver is able to test themselves against maintaining the fine line between safe and challenging. The emphasis of safety is on the protection of those inside the vehicle and does not take into account, except where standards demand it, the impact on others outside the vehicle. As cars become safer for the driver they must be driven faster and harder to test his ability. The car is seen as drawing something from the (male) driver that is natural and essential to his maleness but at the same time it defines and constitutes his maleness.

The car as an object of consumption has many meanings built into it and attached to it that will be explored further throughout the book. The particular ways in which the car has been articulated have impacted on and been shaped by social relations. The kinds of social relations between cars and drivers as well as between different drivers generated in and through the car have been explored here through the idea of articulation, while taking up Dant's point, that it is the assemblage of car-driver that is articulated, through an embodiment the car helps to express and facilitate.

The car affords expressions of masculinity and femininity as well as class and more particular identifications. A young man who drives a ute gives out different signals about himself to those who drive a hotted up sedan like the WRX or the

**Figure 1.9   Holden SS Commodore. High performance and masculinity define
              Australia's most popular car**

older style Commodores and Falcons popular with young men. Perhaps the ute says
something about the young man's physicality or ability and willingness to take on
physical tasks while the WRX suggests an untamed wildness and unpredictability.
The sporty versions of new cars like the Falcons, Commodores and Toyota Camrys
say something about the ability of young men to get what they want and to display
themselves as masculine, able to take on the power, to perform and to wear the sporty
image. The car in its various articulations affords the expression of personal image
as well as affording the demonstration of skill and prowess. The car has certainly
proved a valuable affordance in shaping and displaying masculinity in its various
forms.

The affordances and particular dominant articulations of the car have resulted in
the emergence of driving cultures favouring particular driving styles and associating
different groups with different types of cars and driving. Concrete embodied
experience related to cars is important in informing theoretical approaches to mobility
issues but it is also important to examine the broader cultural background of cars and
the appropriation of them into dominant meaning systems, such as that of aggressive
masculinity through notions such as affordance and articulation. The dominant form
of masculinity as it is promoted through cars will be considered in more detail in
later chapters. This chapter has explored the ways in which the car itself has become
rountinised in drawing out the systematic articulations of particular types and models
and the driving styles associated with them. As Dant emphasises, it is the assemblage
of driver-car that has particular meanings and enables the affordances of the car
beyond mobility. The car can be recognised as a shaping influence in itself while also
being shaped with and through dominant social norms and values. The next chapter

will explore boredom and pleasure in driving as they relate to times and places and how these experiences become routinised through socialisation and enculturation in driving and are shaped in relation to prevailing notions of freedom.

Chapter 2

Inscribing Driving: Boredom and Pleasure on the Roads

The second dimension of driving as a cultural practice to be explored in this book concerns experiences of place and time. Commuter driving can be experienced in quite different ways from pleasure driving or cruising, for example. For many of the research participants the experience of driving to work was coded as boring and often as frustrating, due to traffic. Driving for pleasure had the added dimension of driving in different places such as 'the country', or to the beach, and not in the everyday commuter traffic, making the experience of driving enjoyable in itself. Young people's comments suggest they develop significant differences in driving experience between driving to work and driving for pleasure. While most enjoyed driving of any kind when they first got their licences, they have since developed more coded experiences of time and place and taken on the typical 'attitude' to driving for work (necessity) as opposed to driving for pleasure (choice).

Informing and shaping the coding of these experiences are particular cultural ideas of time, space and freedom that are commonly connected to cars, such as the car as time saving device, the sense of spaces as being condensed and expectations of freedom from obstruction. The driving experience is informed by expectations of freedom as absolute such that it is expected that movement in a car should be unhindered and open to the free expression of individual desires, rather than shaped within the social context of traffic, rules and varied desires.

According to Margaret Morse, there is a level of unreality she refers to as a state of distraction involved in the freeway, the mall and television that invokes feelings of both pleasure and boredom. From this state of distraction the freeway is considered 'a 'vast wasteland' and a waste of time as well as part of the dreams of modernity for equality and opportunity. In contrast Tim Edensor considers the commute on the motorway as an experience of engagement and interest, and contests the representations of motorway journeys as 'boring, sterile and alienating'. Considering the state of distraction outlined by Morse and the engagement of driving highlighted by Tim Edinsor (2003) as an alternative to the distracted state, the chapter will show the routinised role of boredom in driving practice as a way of engaging with the social context of the roads. The sociality of driving, it is argued, is denied in the emphasis on individuals isolated in their vehicles. The 'fiction effect' Morse refers to as distraction is a 'partial loss with the here and now' and related to split belief in which one momentarily sinks into another world knowing it is not real (Morse 1990, 194–194). There is a social complicity in this knowledge of unreality that is played for its social value and as a form of engagement with the roads. The play between fantasy and reality, an ideal world and the real one are mediated by emotions that act within the social formation of driving as a practice and encompass articulations of cars and places. The chapter will explore the differentiations of time and place in young people's comments as well as the articulation and symbolisation of roads through their cultural history and signage.

Driving time

The mobile and home centred way of life in Western societies usefully referred to as 'mobile privatisation' by Raymond Williams (1975) is central to the routinisation of driving places and times. The separation of home and work and the need to move between the two, usually in private cars, has arguably contributed to a sense of distraction seen by Morse as a defining characteristic of 'mobile privatisation'. Automobility has been a quintessential characteristic of modernity along with the cultivation of freedom and the boredom generated by it. Automobility offered a sense of control and domination of distance and time on a personal level, through the car. The car altered the sense of time and place condemning its users to flexibility and a very personal and particular sense of time:

> Automobility ... coerces people into an intense flexibility. It forces people to juggle tiny fragments of time so as to deal with the temporal and spatial constraints that it itself generates. Automobility is a Frankenstein-created monster, extending the individual into realms of freedom and flexibility whereby inhabiting the car can be positively viewed, but also constraining car 'users' to live their lives in spatially-stretched and time-compressed ways. The car, one might suggest, is more literally Weber's 'iron cage' of modernity, motorised, moving and privatised. (Urry 2006, 20)

The 'instantaneous time' of automobility has to be managed in 'highly complex, heterogeneous and uncertain ways' involving an individualistic timetabling of multiple instants or fragments of time through personalised subjective temporalities (Urry 2006, 20). This sense of time contrasts with the official and objective railway timetable that preceded the advent of the car. Since greater distances can be covered much more has to be packed into the day and the car becomes the standard of flexibility by which other forms of mobility are judged inflexible and inconvenient (Urry 2006, 20).

Time has to be used rather than wasted in modernity. There is the sense of delight in being able to 'save' some time and commonly frustration when more time is taken than was expected, to achieve something. The car promises time saving but increasingly offers time 'wasting', particularly in urban contexts. Small fragments of time are pursued in using speed to gather time on one's side. Young people refer to the frustration of being behind a slower driver and will often exaggerate the time lost and the slower speed of the driver in front. The time gained by speeding is also often exaggerated as the real gains are tempered by traffic lights, slowing down to negotiate bends and corners and other unforseen impediments to constant movement. Young people have reported for example that 'if you drive twice as fast it takes half as much time' claiming a freedom and flexibility available through the car that is unreal. It is rarely the case that one can drive at twice the speed in half the time even on a motorway, as it is necessary to slow down for entering and leaving traffic, dense traffic, other vehicles changing lanes and so on.

Figure 2.1   Paris bridge traffic. The reality of driving in the city and urban areas is more stop and start than free flowing movement

The experience of time is changed by the car and part of the experience is 'fantasy time', the sense that time is suspended when we are immersed in what we are doing and the miles are eaten up without us noticing, leading to the idea that cars bring about, not just time savings but a suspension of time. When the time is noticeable and seems to go by slowly, boredom is experienced. This experience is generated by the expectation that the car removes the sense of time being 'lost' by having to travel from one place to another. The car creates the expectation that travel should literally take 'no time at all'.

Driving boredom and pleasure

The differences between commuter driving and pleasure driving or 'cruising' were expressed in the focus groups in comments on driving to work in contrast to driving on the weekend. For many of the participants the experience of driving to work was coded as boring and often as frustrating, due to traffic or the sameness and necessity of the journey. Both young men and women noted the difference between weekend and weekday driving. The weekend represented the opportunity to 'go and explore and just start enjoying it':

> Yeah. During the week you get so sick of it, it just drives you nuts. I've got to drive to town, half an hour each way, and it just sends me nuts. I hate it. So then, when the weekend comes, I love it [driving] again. (Male 2, Goulburn 2)

Driving for pleasure had distinctly different meanings that tapped into popular ideas of freedom. Young people's comments suggest that in the development of their driving experience significant differences between driving to work and driving for pleasure also develop:

> It depends where you're driving. If you're in bumper-to-bumper traffic (laughs) then I don't like driving! But if you're out of the city, out in the country somewhere, it's quite enjoyable. So it just depends ... and who you're with as well. If you're by yourself, it's not so bad, but if you've got rowdy backseat passengers or something like that, then it's not as good. But then, their company can also be good, as well. So it just depends on who you're with and where you're driving, I guess. (Female 2, Blacktown, 21 years).

For a younger man in the group even the traffic could be enjoyable since for him it offered personal space, the company of others and an opportunity to look at other cars on the road. For this young man, who only recently obtained his licence and a car of his own, driving everywhere was a pleasure.

Many of the focus group participants, including women, noted that they really enjoyed driving of any kind when they first got their licences:

> I used to, when I first got my licence it was more of a novelty. But driving as much as I do, I just find it more something you have to do, that you can't really get around. I don't not enjoy it, so if I had the choice of taking ten minutes to get to work rather than fifty minutes, obviously I'd rather drive the ten minutes each day. (Female 1, Blacktown)

The difference between driving to work and driving for pleasure became very pronounced for many of the young people who had been driving for more than a year or two and some driving experiences were consequently classified as 'a waste of time'.

The traffic as well as time could be a factor but ultimately freedom from the restraints of time and 'somewhere to be' were more often associated with enjoyment: 'You're just on the road, no-one's in front of you and you can just go and take your time. But if you've got to be somewhere, at a certain place, then time's important'. (Male 1, Fairfield) Here, the imagined pleasure of the open road contrasts with the pressure of other vehicles, as well as having to be somewhere. Time is seen as constraining and the car is the ideal vehicle of freedom. Freedom is the absence of time constraints, of other vehicles in front of you, or having to be anywhere so you can just go at your own pace but this is a very rare event and appears as a denial of sociality. The car is a means of experiencing freedom in a vacuum where there is nowhere to be, no demands from anyone, whether family, friends or an anonymous other driver thus lacking in any sense of sociality. At the same time sociality is necessary to the facilitation of freedom enabled by the car – the freedom to be somewhere, to visit someone, to develop relationships whenever one wants relies on a social structure and network of roads, regulations and ongoing technological development as the volume of vehicles increases.

Figure 2.2   An ideal road in the English countryside – challenging corners and freedom from constraints

Boring by nature

It is commonly agreed that motorway driving, commuting to work, driving in traffic and on long stretches of highway are experienced as boring. It is variously considered when people's comments are investigated, that it is the road that is boring, driving in itself is boring or the circumstances of the drive make it boring. The following will discuss the various responses from young people that entailed driving being experienced as boring in some way. These experiences will be contrasted with those experiences that were considered pleasurable. For some young people, especially women, driving is just boring due to what to them amounted to a repetitive activity that involved a great deal of sameness:

> I get bored. Like, once I get past ten minutes of driving I'm just like … ugh. Not because it's hard, it's just boring. It's the same stretch of road for ages. Especially driving home, it's so boring. (Female 3, Lawson)

> Yeah, by the end of the week … I don't know, I just have always lived a long way from work, so I never really want to drive on the weekend, just go for a drive … (Female 1, Shellharbour)

The lack of engagement with cars and driving traditionally associated with women seems to be confirmed here. Those men and women who do not like the experience of driving in itself see it on one level as merely a necessary way of getting where they want to go. The fact that they are able to get themselves where they want to be seems to be lost in the mundane task of having to drive. These young women are referring to long distances that are greater than those travelled by the average commuter due to living outside metropolitan areas, however, they are unable to accept the fact of the distance and the sameness of the road as if it should not be that way.

Confirmation that it is the road rather than the activity of driving itself that makes the drive boring can be found amongst other drivers. As one young woman relates, her uncle who is a professional driver, concurs:

> It gets a bit tedious driving back and forth from Sydney or Campbelltown. The road's pretty boring, and even my uncle agrees with that because he was a truck driver and did the highway. (Female 2, Goulburn 1)

The verification that the road is boring for her uncle gives a social status to the branding of the drive along the motorway as boring. The drive is perceived as inherently boring due to the nature of the road and not just the task of driving. The new motorway contrasts negatively with the old road even though it saves time. The more demanding winding old roads, a source of frustration and complaint before the advent of the motorway, is romantically regarded as less boring. This could be considered a result of lack of engagement and removal from the context as Morse (1990) has argued or lack of positive social meaning.

For some young men speed more commonly provides a means of distraction from boredom through thrill and excitement:

> It sort of gives you a bit of a distraction when you're going fast, as well. When you're driving on long trips it sort of gets boring, so if you can you try and have something to keep you occupied. (Male 2, Goulburn 1)

More has to be offered to make the drive a thrill and a source of excitement, and in this the men concur with the women that some roads and some trips are implicitly boring, however they distinguish themselves in finding thrill through extending their driving ability. They want to emphasise that it is not the task of driving in itself that is boring. The complexities of gender in the practice of driving are appearing here. For some of the young men driving itself becomes a means of distraction even allowing more time to be consumed in avoiding boredom through emphasising and exercising their skill:

> I'll take every back road that I can, the old Yass Highway, go round through Yass and all that because I get bored otherwise. I like exploring. Take me time! I think that's why it takes me five hours! (laughter) It would only take me four otherwise! (Male 2, Goulburn 2)

> I sometimes get bored and just go for a drive around. See if I see anyone. (Male 5, Wagga Wagga)

> It takes your mind off things when you drive. (Male 3, Fairfield)

Men are traditionally permitted to find more enjoyment in their mastery of the machine and pushing it beyond the 'allowed' boundaries. The type of car can thus be the means of excitement for the men, a break from the boring routine of having to do something that 'wastes your time':

> I think when I've got somewhere to go, like to work and uni, and I'm on my own time, I hate driving because it's just annoying. It's just something that you have to do to get there, and it wastes your time. I'd rather be doing something else. But then, if I'm at work, like I said before, and I jump in a brand new Commodore with sequential gears I love it. The power and ... I don't know. It's just nice. It's nice to get out on the road. (Male 1, Shellharbour)

Going fast and having a powerful car to drive gave the young men a break from the apparent boredom of their everyday activities. Wasted time is his own time having to be spent on getting to work or uni. There is thus an ambiguity in driving time as an avoidance of boredom and as a boring waste of time that is also evident in this statement:

> I don't get my kicks out of it, but I don't hate it. Like, I guess I like having music on when I'm driving, because I'm a big music buff, so I like that part of driving. That aspect where you can sort of forget about the other things and listen to music and cruise along ... but no, I don't necessarily go out, as we were saying before, go out for a drive just to get in the car and drive. (Male 1, Shellharbour)

It was less likely that males would say that they just did not like driving whereas for women this was more common. For the men it was apparent that the boredom had to be related to something about the driving such as the constriction of having to drive to work or drive on a road that was lacking in challenge and excitement. The women rarely found driving of interest in itself because they did not tend to associate themselves with the car and the skilled performance of their driving. Many mentioned the independence that driving gave them and would refer to driving as something they did so that they could choose when and where they went:

> Driving isn't one of my favourite things to do. I mean, I drive because OK, I have to go to work. I drive on the weekends because I want to get to the place I'm going. I usually tend to drive whether or not someone else is so that I can leave early, but if I know that I'm going to a place where everyone's going, everyone's planned that we're all going to go together, then I do like someone else to drive. (Female 6, Bankstown)

For the women, not expressing a desire to drive was more appropriate while the men were more likely to feel the need to express their enjoyment of driving in some form. Some women did nevertheless find driving enjoyable, relishing the freedom and the potential for adventure the car offered them:

> I love just driving for the sake of driving. (Female 4, Lawson)

> I'd quite happily take a road trip all by myself, with the baby in the back. (Female 5, Lawson)

While the independence the car afforded them was appreciated this often became lost in the routinisation of driving as a cultural practice and all the implications that are involved in that, which include implicit meanings that go beyond the realities of what the car is able to offer.

Space to oneself

While it was commonly believed that driving to work was necessarily boring and that driving on the weekend when your time is your own was a very different thing, it could also be recognised by both men and women that the drive to work itself could be meaningful:

> I like driving by myself. It does give you time to think. You can go at your own pace, look at stuff you want to look at. (Male 2, Goulburn 2)

> I enjoy driving. Most of my driving's done by myself. I do a lot of driving with work. On the weekends, I like to go and visit family and they're in Bathurst, so it's a two-hour trip. I just like driving and listening to music. I find I reach a sort of point in my journeys where I just really take in my surroundings, 'it's good to be back' sort of thing. (Female 1, Blacktown)

For the young man however, driving by himself meant being unrestrained by the desires of others and doing as he wants:

> If you're with someone else, they usually want to be somewhere by some time, so you have to rush to get there or whatever. If it's you, you can stop off by the side of the road and have a bit of a look at this and that on the way. Take your own pace. It's good doing it that way. (Male 2, Goulburn 2)

Other young men expressed similar sentiments:

> Like, I've got my CD's and I can make it as messy or as clean as I want and no-one can say anything because it's my car. And you can just do whatever you want! (Male 2, Shellharbour)

The sense of not being constrained by others in any way is central to these experiences. While the car is often an emblem of sociability for young people affording opportunities to meet up with others and go places they could not get to otherwise, as well as to participate in a culture that signifies adulthood, this is tempered by the sense of individuality and freedom that the car also expresses. The 'mobile sociability' offered by the car is overshadowed by the emphasis on individual and independent freedom of movement often illustrated in advertising, and the fundamentally social nature of driving is lost.

Comfort and familiarity

A distinct sense in which place and time could influence driving practice was in the level of comfort and familiarity with the car, the place and the time. Driving in the city and driving at night were experienced differently according to familiarity with city and night driving and comfort with the car. There were varied responses to these experiences, with some being quite used to driving in the city and comfortable with it though they found it inconvenient, while others did not like it at all:

> I don't like driving in the city, because I don't do it a lot particularly, and I just sort of find it a bit intimidating, but when I know where I'm going, I'm a lot more comfortable to when I'm going somewhere I've never been before. (Female 3, Blacktown, 22 years)

A certain level of comfort and familiarity are part of the routinisation of driving practice but cities and traffic in particular, are often experienced as intimidating, as comments from many young people indicated.

A young woman from Goulburn, a regional town of New South Wales, would not drive in the city, 'no way', and others from Bankstown in Western Sydney found taxis more convenient when they went to the city. A young man from the Fairfield group also in Western Sydney did not like driving in the city at night whereas during the daytime it was a different matter.

The context and surroundings can be an influence in other ways such as the slow speed appropriate for 'lapping' around the streets in specific locations such as the Rocks or Brighton, in contrast to the displays of fast acceleration, lane changing and speed on highways and motorways.

The comfort of cars is another aspect that is of primary importance and has been an advertised feature since the invention of cars. Cars have become very comfortable for their occupants to the extent that they can be considered 'armchairs on wheels'. The level of comfort induces relaxation and a sense of security for the driver and passengers that promotes a state of distraction according to Morse (1990) where the environment of the car is more real than the world outside and the driver is in an idealised world separated from the outside world. The driver is capsuled in an idyllic world with controlled climate and selected sound in an intensely private space:

> ... lifted out of the social world, the driver is subject, more real and present to himself or herself than the miniatures or the patterns of lights beyond the glass, or farther yet, beyond the freeway. (Morse 1990, 199)

Comfort and familiarity appeared to be a significant part of the routinisation of driving practice. Many of the young people thought about their driving from the perspective of their own familiarity with their car and their surroundings: 'You get to that point where you've been driving it so long that you just know every inch of it ...' (Wollongong, 20 years). Their awareness of social space, however, did not always extend much beyond their own vehicle to other road users.

Disengagement from the world outside accompanies the 'intensely private' comfort of the inside of the car (Morse 1990, 203). This is significant for many young men who see the road as a site for their peer sociality, but not necessarily as a space involving interactions with others beyond their peers. One young man described his driving from the perspective of his mother, contrasting it with his own sense of control: 'My mum's constantly on my back, telling me to slow down ... she thinks I drive too fast, but really, I just drive to what I feel comfortable.' (Goulburn)

This young man focused on whether he was driving fast or slow enough for his own comfort and not whether he was driving to speed limits or traffic conditions. He insisted that being too concerned with speed could have consequences for concentration, suggesting that the real problem was distraction, not speed, and that he knew what speed he could handle. He stated, 'if you worry too much about the speed and where you're going, it's taking your mind off other things and you end up having a crash'. This statement highlights an incongruity between the desire to maintain the idea of freedom the car is meant to embody, and the restrictions of the actual driving experience. Admitting to a lapse in concentration appeared less threatening to his sense of control than acknowledging that his choice of speed could be questionable. For many young men demonstrating their ability to handle speed is an important part of their social development and their sense of control.

Comfort and familiarity from the driver's point of view then, are significant factors in how and when the young people will drive. Some of the young women also noted the significance of familiarity with their car and their surroundings in how they drove. The deeper meanings of this familiarity and the extent of social involvement in the creation of meanings of familiarity and boredom and pleasure have been explored in the preceding sections. How places are socially performed and articulated will be considered in the following.

Performance places

Places as well as vehicles, are articulated through driving practices which encompass the implications of the place, and the associations and attachments to it expressed in discourse. Experiences of the road are intertwined with discourses and aspects of social and cultural practices through which it is articulated (Redshaw 2007). Vehicle type and characteristics of the driver produce articulations of cars that help to shape the experience of the road. There are expectations requiring particular performances if driving roads with a reputation, such as the Great Ocean Road on the coastline of southern Victoria in a sports car or riding a motorcycle. Campervans and other 'tourist' vehicles carry different expectations and demand different performances, often the impatience directed at the slow meandering pace of the tourist. Vehicle type denotes particular driving expectations or articulations demonstrated in performance for both those inside the vehicle, and those in other vehicles as does the context of the road itself.

For Michel de Certeau, well known for his account of 'Walking in the City' (1984), space is a performed place, place having a stability and ordering that implicate specific meanings within it accompanied by expected ways of acting, and practices that are appropriate to it, including those that might be outside certain ideas of etiquette or law. Space on the other hand, is constituted through movements and velocities that make it a particular defined space (Urry 2000, 53). There are meanings implicated in places and spaces then, and the actions or practices that are cultivated within them.

Marc Augé's (1995) exploration of the motorway phenomenon cast motorways as non-places that produce detachment and decontextualisation. In non-places users are unable to recognise their own presence so that the individual becomes 'frozen', 'merely a gaze', only able to experience and engage with their surroundings through the mediation of words and texts (94). Merriman in his critique of Augé maintains that the newness and difference of experience claimed for supermodernity is too extreme and that the discontinuity between traveller and landscape could also be said about railway travellers in the 19th century (Merriman 2005, 150). He also objects to the lack of sociality and subsequent solitariness of travellers in Auge's account and sees this as overlooking the complexity, materiality and heterogeneity of social networks taking place in airports, supermarkets and motorway service areas (151):

> Airports are spaces where travellers *may* experience feelings of boredom, frustration, solitariness or dislocation, but these experiences are neither inevitable nor limited to such spaces of travel and exchange. Academics must not overlook the complex processes through which such feelings are enacted, and ... pay more attention to the complex histories, geographies and sociologies of such spaces. (152)

It has been argued by others that different forms of travel from horse-drawn carriage to railway to motor vehicle resulted in new forms of attachment rather than detachment, and that it is not the form of vehicle that produces disengagement but control of one's own pace (Schnapp 1999). Being a driver of a car produced a different kind of control than that experienced in other forms, it is seen as more active whereas being a passenger is considered merely passive. Characteristics such as place and time are not taken into account since it is the mastery of speed that is of most interest in Schnapp's historical account. Having control of one's own pace is a popular idea of the car that tends to be pitched against social control. This certainly produced new forms of attachment in the experience of mobility through the car and it is these new forms of attachment and their underlying sociality that has contributed to the place and importance of boredom in driving discourse.

The diverse mobilities that characterise societies into the new century include movements of peoples, objects, images, information and waste (Urry 2000, 1). Mobility is thus a central defining feature of globalisation and the crossing of national boundaries that typifies it. In new ideas of place and how places are lived, places are understood less as fixed in their meanings and attachments:

> ... movements of travellers are not seen to be movements across the landscape; rather these flows and associated frictions and turbulences are integral to the construction and performance of landscapes and places. (Merriman 2005, 146)

Merriman points out that there are few critical accounts of 'the geographies, sociologies and anthropologies of *driving along* specific roads or through specific landscapes' with the exception of Edensor's (2003) M1 experiences and Williamson's (2003) account of Malaysia's National Expressway (146). The accounts that he is looking for are those that relate concrete experiences of places and how they are lived. Route 66 in the United States is an example of a road that has very specific historical and cultural meanings that could potentially dictate how the route is driven and in what kind of vehicle for the ultimate experience. Such routes are articulated through the meanings applied to them including the type of person and vehicle considered most suitable to the full experience of the route.

Country roads and back roads

> The back road's the best, obviously, because it's not boring. You've got plenty of corners. (Male 2, Goulburn 1)

Roads then are articulated in the sense that cars are with particular roads having meanings related to the driving practices they are historically connected with. Four lane highways are routes to and from places and not necessarily enjoyed for the drive or scenery and yet they are seen as the solution to traffic problems and as facilitating speed. Particular roads offer spectacular experiences of driving or riding such as the Great Ocean Road in Southern Victoria or the old Pacific Highway winding through the dramatic coastal ridges and valleys north of Sydney. Roads are deeply inscribed with mobile meanings that amount to articulations of the road. Articulation in this sense is intended to invoke the 'aliveness' of place and the performances that are implicitly connected to it. How the road is articulated is related to its surroundings as well as to the practices entwined with it.

Traditionally, back roads in Australia, roads that are not the major routes, are escape routes from the traffic, the police and the rules of the road. Documentation on when speed limits were applied to many roads in Australia is not easy to come by but it is well known that there were no speed limits on many roads until the last decades of the previous century when a maximum of 110kms/h was brought in across the country. This applied to most states in Australia except the Northern Territory where a maximum of 130kms/h was brought in only in 2007. The Territory was seen as the last frontier where a car could really be opened up to speeds of 200kms/h and over. The higher road toll prompted the Labour Chief Minister to finally bring in a maximum limit, which the more conservative Liberal and National governments resisted. Motorways became the only roads on which speeds of 110kms/h could be achieved in New South Wales in the 1990s.

Back roads through suburban streets often colloquially known as 'rat runs' provide an alternative to the traffic queues but tend to result in drivers speeding on suburban streets in order to maintain the speed they felt they should have been able to travel at on the main routes. Main roads might allow speeds up to 80kms/h but suburban limits have now been lowered from 60kms/h to 50kms/h. Nevertheless drivers strive to get ahead of where they might be if they had stayed in the traffic by exceeding speed limits to make it worthwhile taking the back roads.

Getting through the traffic as fast as possible overrides any sense of the local community and the costs and risk to communities of speeding on suburban streets. Communities have fought back and pressured local councils to employ methods to decrease traffic speeds. Local governments have responded by putting roundabouts, speeds humps and chicanes on local streets to slow the traffic down and putting more traffic lights and stop signs to discourage use of back roads and increase local traffic flow.

Escaping down the road

The back road coincides with the dream of the open road and the true sense of freedom where there is more opportunity for individual expression of driving prowess and of being out there on one's own and not merely part of the mundane flow of highway traffic. The social meanings underlying the contrast between the flow of traffic on the highway and the freedom of the back roads involve the association of passivity with being part of the flow and adventure and agency with finding one's way around and getting ahead of the flow:

> Going down the road, symbolised not only a way out, a going to and a getting away from, it represented possibility, risk and romance. The road and at the same time, those who lived according to its rules, the throw of the dice, the chance of a new start and the ever present danger of failure and even death on the unknown highway, were invested with all the symbolic power that the frontier and the frontier's men carried for earlier generations. (Eyerman and Löfgren 1995, 57)

Road movies illustrate the themes of fantasy and escape that the car is expected to enable. Films with road journeys are generally about freedom and mobility and have been considered in two major themes: the escape motive and the quest motive (Venkatasawmy, Simpson and Visosevic 2001). According to Eyerman and Löfgren (1995) discussing images of hitting the road such as those found in road movies, can reveal something of the rhetoric and the experience of social mobility. They propose that films about the road and the journey present a moral discourse that is a reflection of the society they emanate from (60). With key themes of freedom, escape, danger, risk and some form of excessive attachment to the automobile, the films typically depict either normal life as negative against the adventure and freedom from ties and responsibilities of the road, or the environment of the road as inherently hostile, risky and dangerous and normality as safe, secure, and at least potentially meaningful.

The road journey is typically a journey for men and their machines into a realm where anything can happen, where anomie and anarchy threaten and the darker side of human behaviour can emerge. The idea that 'it's a jungle out there' is often represented in movies (Eyerman and Löfgren 1995, 67), for example the *Mad Max* series. The nostalgia of the road for a lost freedom and the dreamlike quality invested in the car are both played out in the road movie even if the freedom is ultimately one that involves chaos, violence and continual threat. The appeal of the open road that the road movie draws on is the potential of arriving somewhere new, escaping what is familiar, boring and determined, and experiencing the excitement of adventure, venturing into the unknown, '... the freedom to move on, to make one's own way and to risk everything for a new chance' (Eyerman and Löfgren 1995, 74).

Road movies draw on the fantasy of the road as liberation from responsibilities and restrictions of relationships repeating themes of escaping the meaninglessness of existence. The road is enticing because it represents the wild unknown as well as the potential for liberating experiences recreating a broad but nevertheless deep articulation of the roads as a source of freedom whether as liberation through discovery or chaos.

The drama of the road

> driving is a culturally bound procedure organised around which manoeuvres, forms of etiquette and gestures of annoyance, for example, are 'proper' in particular contexts. (Edinsor, 2004: 112)

Driving the Great Ocean Road in southwestern Australia is one of those exhilarating, spectacular experiences that apparently begs fast paced handling of powerful machines, motorcycles screaming around corners and testing the metal of sports cars. It is the road on which to take a Ferrari for a run and sports cars can be hired form Melbourne for such purposes. The posted speed limit on parts of the road of 100kms per hour seems far-fetched given the narrow lanes and very tight bends. There are only two lanes, one each way for most of the road, giving rise to the question of how the speed limit has been determined. As one approaches a fierce bend in the road, on either side of the road with an ocean view to die for, in front are the standard speed limit signs indicating 100kms/h. Driving with caution allowed about 40–50kms at most, around those bends, of which there are many. There were few stretches where one could realistically consider getting close to 100kms. The speed limit has been lowered to a maximum of 80kms/h amid significant controversy played out largely in Victorian newspapers. The significance of the road as a place for the performance of driving mastery is well known.

With tourist traffic on a long weekend, holidays or any weekend for that matter, speed on the Great Ocean Road becomes concerning. Those wishing to try out the prowess of their driving and their machines compete for space on an extremely narrow road with those unfamiliar with the road wishing to look at the spectacular scenery. While I was there, an accident occurred which involved a car swiping a van going in the opposite direction and then going off the road and over a cliff, narrowly missing fishermen below. The car contained a young couple with a baby, and the accident was reported in the local newspaper. All survived the crash unharmed, fortunately, but the incident shows that not much has to go wrong for there to be some serious consequences. At least two people a year die on the 245 kilometre stretch of road,

Figure 2.3   Sports car, speed and challenging bends on the Great Ocean Road

and dozens are seriously injured. In the years 2001–2005 twelve people were killed, all male, and 167 seriously injured (VicRoads). Motorcyclists are particularly over represented in the crash statistics (Great Ocean Road Region Strategy 2001).

The road is a narrow, winding, dramatic road for much of its 245 kilometres. Blind tight curves are common in the coastal sections and though there are two clear lanes, there are almost no overtaking lanes. Speeds drop to 80kms/h then 50kms/h through townships occurring roughly every 20 kilometres. A variety of vehicles are found traveling the road from buses and trucks to all kinds of cars, vans, motorcycles, and a range of camper vehicles.

A website for Tourism Victoria describes the attractive features of the road:

> The panoramic views of the Great Ocean Road are breathtaking. The road takes you along cliff edges, around hilly slopes, down onto the edge of beaches, across river estuaries and up to breathtaking headlands.

Tourism World Magazine's website gives a romantic quality to the historic account also noting the drama of the road:

> The road was mainly hacked out of the earth by manual labour and was completed in 1932. It is a long and winding road that starts officially at Torquay and finishes at Warrnambool.

> The problem with it is that although the views are almost constantly breathtaking the driver of a car sees little of them because this road requires your full concentration. The best way of dealing with this is to pull over at most if not all of the viewing stations and parking bays along the way. Otherwise only the passengers will enjoy the views which are, indeed, amazing.

> (worldtourism.com.au/australia/vol1_issue1/ accessed August 2003)

Figure 2.4   Great Ocean Road entrance

One way in which the road is articulated is through the signs that are posted along its length giving warnings, appeals and directions to drivers. The signs indicate the preferred mode of travel on the road or the appropriate road etiquette. Many of the signs are designed for visitors who do not know the road. There are many parking bays along the road for sightseers labeled 'Slow Vehicle Turnouts' accompanied by messages encouraging drivers to 'consider those following'. Warnings such as 'Keep left unless overtaking' are common on Australian roads, but these signs are specific to an area that has a highly mobile tourist population.

There are many signs warning of the dangers of the road and particular accident zones. One sign reminds overseas visitors that we drive on the left in Australia with a simple representation of two lanes with arrows indicating direction. High accident zone signs as shown in Figure 2.6 appear fairly regularly along the Great Ocean Road, but travelling close to the vehicle in front is also quite common.

There are no signs warning of the dangers of close following, however. The emphasis of the messages that are present and the absence of messages addressing other problematic behaviours is significant and gives a particular articulation to the road. The messages do not ask that those following take care not to follow too closely, become impatient or expect everyone to be driving as fast as they might like. There are no signs asking for consideration to be given to sightseers who do not know the road. The discourse of the road appears to favour those who travel the road at maximum speeds as a means to get somewhere, who are aiming for a particular destination or for the experience of driving the road as fast as possible.

Distances and times to reach towns along the road are given in the signs shown in Figure 2.7 indicate to drivers how much driving is involved. Drivers are encouraged to 'plan your trip' so as to allow a more fluid and well-paced journey with planned stops. The distances may be surprising to many overseas visitors who are not used to the vast spaces typical of Australia. Tourist stops for scenic opportunities, known as

Figure 2.5   Consider others following – signage on the Great Ocean Road

viewing stations and parking bays, are also being flagged as places to aim for. These are typical of the well-known scenic sights such as the Twelve Apostles where large car parks have been constructed to accommodate increasing numbers of vehicles.

The extent to which values of speed, competition and the exhilaration of the machine have dictated driving speeds is apparent in the messages conveyed in road signs. The emphasis on slower drivers moving out of the way of faster drivers is potentially inflammatory, especially in the absence of signs addressing the aggression of those expecting priority. The road appears to be articulated for maximising speed rather than simply for access and enjoyment within reasonable safety margins even though it is clear that tourism, in the form of exploring the landscape is the dominant activity on the road.

With new improved road surfaces and the addition of the occasional overtaking lanes, expectations of greater speed are very high on many roads. There has been considerable emphasis on increasing speed, and governments in the past have bought into and encouraged the perception that greater speeds are possible. Advertising of new roads and motorways, improvements made to old roads, lobbying by motoring organisations and a concern on the part of governments with letting the public know they are doing what they can to streamline car travel, have contributed to the expectation of greater speed (Davison 2004). Governments have certainly thought greater speeds were desirable, although more emphasis is being given to other factors in recent decades such as the impact of urban speeds, especially on children.

Road safety discourse and thrill/speed discourses have been around for some time – since horse and carriage days – according to Jeffery Schnapp's (1999) analysis of

Figure 2.6   Accident zone and two way warning signs

literature of the nineteenth century. It is interesting then to find that most mentions of the Great Ocean Road refer to the scenery to be enjoyed on the road suggesting that it is the view from the road, not the road itself that is of primary interest. And yet the 'GO' Road is legendary for the experience it offers of exhilarating driving and riding. While road safety discourses have been around for some time, they have been dominated or at least moderated, by speed and thrill discourses. The proposed lowering of the speed on the Great Ocean Road from 100kms/h to 80kms/h represents a move ahead on the road safety front. Perhaps it is indicative, as is the lowering of speed limits in other areas, of road safety overtaking thrill and speed discourses for the first time since the 1920s.

Speed of motor vehicles has increased significantly in the 20th century with limits in most of Australia now at a maximum of 110kms/h. Meanwhile manufacturers claim speeds over 200km/h for some vehicles and urban speed limits have been reduced to 50kms/h in many parts of Australia with quite a lot of support from the community. Speeds have also been reduced on motorways. The Geelong freeway was 110kms until it was upgraded and improved, at which time it was reduced to 100kms. There have certainly been objections to this, as with other motorways, since motorways raise the expectation of higher speeds.

It is interesting however, that speed discourses are almost absent from public arenas publicising roads such as the Great Ocean Road, apart from letters to newspaper editors and the specialist magazines. Tourism has a victorious presence amongst the web pages and pamphlets referring to the Great Ocean Road and these generally recommend the scenic aspects of the road requiring a slower meandering pace. Many viewing spots have been added along the road to enable tourists to pull over and enjoy the view without disrupting traffic by moving slowly on the road.

Speed discourses belong to an implicit agreement that speed is valid and to be prized and is therefore left out of the field of public discourse where it could be debated. Although tourism is the predominant mode of travel on the road the signs

Figure 2.7   Distances and times to destinations on the Great Ocean Road

along the way privilege speed. The experience of the road for the enthusiast is the authentic experience of driving the road and this involves maximising speed in order to pitch car and driver and road against one another. Tourism is regarded as passive in comparison to those seeking the true driving experience. It is the connection between speed and particular ideas of freedom that explains the apparent contradictions that can be read in discourses of the Great Ocean Road.

Habits of freedom

Freedom is one of the abiding characteristics of the car and is frequently mentioned as a value that is prized. Freedom in this sense often means the freedom to go where you want, when you want and with whomever you want. Young people mentioned not having to wait for a bus or train or anyone else, whether they were at home or at a party, with their own car they could leave when they chose. Not having to rely on parents or friends was a big factor for both young men and women and being able to 'get places'. Some considered the advantages of the car very loosely as: 'Just the freedom, basically, to do what you want' (Male 4, Goulburn 1) and others gave very detailed accounts of having to be without their car when it was being repaired or they lost their licence. One male described the 'foothold to independence' achieved by learning to drive as 'part of growing up' that made him feel he was 'on track' (Male 1, Shellharbour). A woman talked about the excitement of being able to go to new places: 'I think I drove for about four hours the day I got my P's! (laughter) Every street! And you're like, wow! I've never seen this street before! I was very excited …' (Female 1, Goulburn 2). Another young man described becoming 'a convenient and independent person' (Male 5, Wagga Wagga).

Freedom on the roads means much more than this however. Car manufacturers tend to promote a form of freedom that is absolute, as if there were no restrictions, such as the need for fuel and the realities of traffic. Many high-powered cars are advertised on open and dramatic roads with a male driver driven by his apparently implicit desire for speed and the thrill of potential loss of control (Redshaw 2007). Manufacturers claim to be appealing to fantasy in their advertising however they continue to promote the need and desire for faster, more powerful cars when the reality of speed is that optimum levels that can reasonably be expected to be managed with growing numbers of cars have already been reached. Higher speeds clearly produce higher fatality and injury rates. There is an element of fantasy in the creation of reality and part of the background to ideas of freedom in cars lies in the broader social context of ideas of freedom and how freedom is used in the so called 'free world'.

The feelings of exhilaration accompanying experiences of speed are often associated with freedom and this is typically regarded as a feeling of freedom from restraint, of breaking the boundaries and exceeding bodily limits. Advertising attempts to draw on this experience showing the road/car/driver experience as one of exhilaration – provided the road is a deserted but challenging one with lots of bends to negotiate. Ford finished many of its television ads in Australia in 2004–2005 with the words 'with the all new Ford (Explorer, Escape, etc) there are no boundaries' followed by an explosion in the heavens and the words 'we have ignition'. The Falcon ad showed a man in his late twenties/early thirties driving on an open road with his hand out of the window scooping in the breeze as he would have done as a child. Bursts of energy showing him with his arms outstretched in triumph and exhilaration are interspersed throughout the ad accompanied by the soundtrack 'I don't want the world to change, I like the way it is ... I can't get enough of this'.

Holden similarly showed a man of late twenties to early thirties enjoying the drive on challenging bends. One ad began with a baby in a stroller leaning into the twists and turns with the text 'you've always had a thing for corners' followed by a scene of the car on a tight u-turn bend in a road with dramatic back drop and the grown-up man 'wrestling' with the power of the car and conveying the exhilaration of the challenge. A soundtrack is overlaid with the lyrics 'don't wanna grow up, I wanna get out, hey take me away, I wanna shout out, take me away'. Pure pleasure and exhilaration is illustrated in both these ads as something peculiar to men and naturally following from their sense of mastery of movement and melding them with the car and this is considered to be the ultimate feeling of freedom. Drives on challenging deserted roads are rare for most drivers, most of the time but the fantasy is nevertheless significant in articulating the car.

The Commodore SV8 was advertised on Holden's website where it was claimed that the car had 'awesome power', 'awe-inspiring handling' that was 'a sensation all of its own' going 'beyond the simple thrill of V8 driving' with 'the understated power of knowing that the absolute authority of 250kW is growling under the bonnet' (www.holden.com.au/www.holden/action/modeloverview?modelid=4009& navid=1 Accessed 24.11.05). Pleasure is firmly allied with power and handling and the masculine engine 'growling under the bonnet' and this pleasure is the sensation of freedom, of leaving it all behind, escaping, being taken away from it all.

Freedom has been considered as liberation from bondage, and this is primarily 'the right of the individual to act in any desired way without restraint, the power to do as one likes' (Rose 1999, 62). Freedom is commonly understood and framed as freedom from coercion and as an opposition between individuals and society, with individuals resisting the imposed regulations of authorities and government. The implication sometimes voiced in letters to newspapers for example, is that individuals would be free if it were not for the imposition of arbitrary and meaningless restrictions designed to do nothing more than restrain 'natural' desires. The systematic favouring of some and restriction of others through regulation requires ongoing investigation, however, it is commonly assumed that road rules are unreasonably restrictive and can be bent as individuals see fit to accommodate their skills and abilities. This is part of the socialisation related to the roads and results from the conflict between different ideologies and the prevalence of risk and consumption as a solution to the meaningless of life.

This view of regulation as unreasonably restrictive and the cause of boredom is related to the denial of sociality and emphasis on individualism in social and cultural practices. Individuals are considered free and society restrictive while it is nevertheless social networks that create and enable freedoms such as the mobility of the car. These same social networks also create exclusions and restrictions within them but this is not the dominant focus of concern. Rather it is the generalised idea of freedom unrestricted by social demands that is defended in order to mask inequities. Frequent references to the prevalence and unfairness of speed cameras are made in the media, particularly in current affairs programs. This is a popular issue with the general population since it appeals to the idea of government as big brother unfairly and unnecessarily restricting the freedom of individuals and making them pay for what comes 'naturally'. Stories about the restrictions and difficulties that pedestrians face appear much less frequently indicating the extent of the dominance and implicit connection of the car with freedom. There have always been protests about restrictions of pedestrian access but these have been overridden by motoring interests.

The understanding of freedom in the absolute sense of unrestricted thrill is cultivated in order to keep the market in demand, the new models having appeal through the 'more' that they offer, generally more power and speed (Volti 2004). Comfort and style are also promoted as important features but performance has been the top theme in car advertising according to research in the 1990s. Ferguson, Hardy and Williams (2003) analysed the content of 850 car advertisements from the United States, Europe and Asia, and listed the primary themes according to frequency of appearance. Performance and sales incentive were found to be the dominant themes in 1998 and the previous decade and a half. Power and performance are strongly connected with the thrill of being free and unattached through the ability of marketing to draw on meaningful associations.

Freedom, fantasy and consumption

The desires that are appealed to in advertising are created and channelled within the context of social and cultural priorities. In commodification, which is one facet of the process of consumption as outlined by Silverstone (1994, 123–131), industrial and commercial processes turn artifacts, objects which have social meaning, into commodities of value and ideological processes are inscribed in them defining them as products that express dominant social values. Advertising and marketing appeals to the need for social meaning by constructing commodities as objects of desire but this desire is insatiable, as goods address specific functions rather than the broader need for social meaning which they purport to fulfill through imaginative and sensational fantasies (Carrabine and Longhurst 2002, 187).

When objects of consumption are taken into the household there is a tension between dependence on 'the material and symbolic objects of mass production' and the expression of freedom as creative participants in mass culture (Carrabine and Longhurst 2002, 188). The object is appropriated in the household's possession and ownership of it but it is then objectified in a way that displays it as expressing the 'principles that inform a household's sense of itself and its place in the world' (188). Incorporated into the moral economy of the household the object may be used in ways not imagined by the manufacturer but which allow it to become part of the routines of everyday life. The object represents both social or shared and individual values.

The relationship between the household and the outside world is understood as one of conversion where some meanings relate to broader public meanings and other more personal and private meanings are not so easily converted. This way of looking at the consumption process 'emphasises the dynamic interchange between the production of popular culture and the ways in which audiences use the objects produced by capitalist enterprise' (Carrabine and Longhurst 2002, 188). Both the fantasy and mundanity of everyday consumption are important here. The way in which an object is consumed may be ordinary, however it is also potentially filled with the fantasy accompanying its articulation through the commodification process.

The ongoing and significant interaction between economic realities and imaginative possibilities referred to by Carrabine and Longhurst (192) is evident in comments from young people about the inconvenience and limits of the car such as having to park it, the costs of maintaining it and dealing with the car when they want to indulge in some drinking. Carrabine and Longhurst want to emphasise the 'extended sociability networks' facilitated by the car and the opportunity for citizenship it brings with it. The sociability that the driver enters into is complex however, and it is important to consider in some detail how sociability and the desire for it brings with it implicit and complex meanings. They highlight the ease with which the car as an object of consumption becomes 'ingrained in the routines of everyday life' in the transition from every opportunity to drive being a 'massive buzz', to the exclusion of all other forms of mobility previously engaged in, such as walking and cycling.

The routine of getting to work can be experienced in such a way that driving is equated with an apparently unavoidable and disliked part of the day. The horror of having to 'wake up' in the traffic on the way to work expressed by one young woman indicates the ambiguity in what the car allows and what it demands:

> Well, I start work early in the morning. Getting in the car is – you just wake up and in half an hour you have to be at work! It's like god! (Female 1, Wollongong)

The car allows the possibility of waking up on the way to work but at the same time the demands of driving make it a rude awakening. There was occasionally some reflection on the sense of being captivated by 'the flux between the ordinary and the extraordinary that animates consumption' (Carrabine and Longhurst 2002, 194) and the predicament this creates for the individual:

> I hate driving, especially around here. If' I've had … not a nice day, and I'm in a rush to get somewhere and there's traffic everywhere, I tend to break road rules. … speeding or something and I've got to get there in an hour. I should leave earlier, but I don't. I leave at the last second, and then it's five minutes and I can't make it, then I'll really try to speed and I get angry at other drivers for driving too slow! (Male 2, Fairfield)

> But isn't it your fault? (Female 3, Fairfield)

> I know! But it's a habit. (Male 2, Fairfield)

The whole process of getting to work and dealing with the traffic in a way that raises his ire is seen as habit and there is a level of helplessness that follows from this. It is 'your own fault' at the same time as it is a habit. The helplessness follows from the extent to which it is the social coding of traffic as frustrating and getting to work in the traffic as an infringement of one's liberties that is governing the practice.

Morse's 'semifiction effect' of distraction, is like split belief but not identical with it – 'knowing a representation is not real, but nevertheless momentarily closing off the here and now and sinking into another world' (Morse 1990, 193). For Morse there are 'two or more objects and levels of attention' combined with 'two or more different, even contradictory, metapsychological effects' in this semifictional effect. There is both a 'plane of the subject in the here and now' and 'the plane of an absent or nonperson in another time' (Morse 1990, 194). The former is related to discourse and the latter to a story.

In the tale of the young man getting himself to work in a state of frustration we can see the here and now exemplified in the reality of the traffic and not enough time, while the absence is an immersion in a world where there is enough time because there is no traffic to interfere with one's progress, illustrated consistently in car advertising. The trip should only take an hour but requires more time in heavy traffic. Peak hour traffic makes the one hour trip an ideal, a story, and yet the 'reality' of the traffic is not accepted.

Alternative engagements

Tim Edinsor argues for the emergence of new forms of sensuality in the interaction between person and machine and the engagement of the body with the machine producing a '"feel" for the road' (Edinsor 2003, 159). The enthusiast, the hoon or boy racer, the tourist and the commuter contrast with each other in their particular feel for the car, the road and their own embodiment. Familiar sensations develop between the distinct sensations of particular vehicles and particular roads so that 'a compendium of sensory experiences [is] built up over time' (Edinsor 2003, 160). Car journeys thus have the potential to be, as he states:

> … full of smells, sounds and tactilities, producing a corporeal sociality that inheres in the intimate relationship between bodies and cars and the spaces through which they move, the distinctive roadscapes, particular models, road textures, and driving conventions and habits.

Edinsor is referring to a motorway journey but it is the agency of the embodied, experiencing commuter he describes that is relevant here. While the motorway appears as the epitomy of the mundane, the 'realm of dull compulsion and habit, boredom and oppression' (155), Edinsor maintains that there are possibilities for transcending the banal. The skillful habits through which interactions with other drivers are metered in accordance with the regulations and cultural norms of the road may tend to minimise reflection. Skillful habits, while part of the 'practical engagement with the world' can become seemingly automatic (155). Nevertheless Edinsor argues that routine practices are 'open ended and fluid', that everyday life rather than being inauthentic and impoverished, has immanent potential for new possibilities.

Habitual experience is complemented by the immanent experience of the everyday including stories, fantasies, disruptions and oddities. While considering driving in traffic as unpleasant a number of young people expressed an enjoyment of the space to themselves that driving allowed:

> I do enjoy driving sometimes, especially Friday afternoons after work, driving home, just when it's not to and from work, and it's that unwind, relax time on the way home from work, generally. Transition time. When I'm driving to work, it's just a chore, really. Especially ... the traffic's always really bad and it's really frustrating. It used to take me half an hour to get to work and now it takes me almost an hour because of the traffic. (Female 2, Shellharbour)

Another young woman had disliked a long distance driving experience but when she came to reflect on it after not having the experience for some time she felt differently about it:

> I used to drive to Canberra every day for work, for just over three years, and I thought I would never do this again in a million years, and I haven't done it. I finished up there in February. Haven't done it for that long, and I went to Canberra on my own on Friday night. And on the way there I had the music turned up loud and I was singing and I was going, 'God, I miss this time on my own in the car! Get another job in Canberra!' Because it was really good. That was the only time I was on my own, really, and I did used to think about a lot of things while I was driving, and I thought I was over it but I didn't realise how much I missed it until I did it again. (Female 3, Goulburn 2)

In this experience the mundane routinisation and its social inscription is seen through. The trip acquired another meaning when it was related to a different context, in this case, going to meet up with people she had worked with as opposed to driving away from home to go away to work. The journey was now bracketed with social engagement.

Ultimately, Edinsor maintains:

> ... mundane journeys are both irrefutably individual and social as they accumulate resonances of our own journeys, spaces, stories and sensations and those of others. ... we might conjecture that constellations of collective experience constitute something akin to a structure of feeling shared by motorway drivers in general and commuters along specific routes more specifically. (155)

While individuals are isolated in their individual vehicles their experience is often backgrounded by a collective level of experience, attributing particular meanings that become routinised. The woman above driving to work in another city, leaving home and family has now established a new network and different feelings about the experience of going to Canberra that is not recognised as such until it is re-experienced in a different way. The routinised background experience that had previously dominated has now been pierced with a more particular experience that has more personal meaning. It is in part the discourse of boredom and its place in the social structure that has contributed to the focus on the isolation rather than the sociality of the individual, but this is also exacerbated by the hostility and conflict between social and individual demands that is characteristic of modern Western societies.

Sara Ahmed discusses the emotions as an interweaving of the individual and the social such that collective feelings are produced. '… Individual and collective bodies surface through the very orientations we take to objects and others' (Ahmed 2004, 39). It is how we come up against others and how emotions are already attached to those connections, that in part determines the feelings we have in particular situations:

> … feelings rehearse associations that are already in place … The impressions we have of others, and the impressions left by others are shaped by histories that stick, at the same time as they generate the boundaries that allow bodies to appear in the present. (Ahmed 2004, 39)

The collective experience referred to by Edinsor as resulting in a shared structure of feeling related to a particular place can be allied with Ahmed's 'collective feelings' and refers to a sociality that is not merely a passive compliance. The sense in which individuals share in the collective experience is not necessarily recognised as an active engagement as we saw with the young man who felt that running late for work and then getting frustrated in the traffic was 'habit', even though how he managed his time and his speed involved his own agency. In considering his tendency to be late and becoming frustrated with the traffic as habit while continuing with this practice, he is engaged with an aspect of sociality that functions as a form of social inclusion which individuals value as social beings. The collective feeling he is immersed in is not an agreement with the social structure and its rules, but with a way of responding to or dealing with those structures and rules.

The social engagement of driving does involve immersion in dominant social perspectives that help to give meaning as well as a sense of shared experience at the same time as individuals are able to structure alternative experiences for themselves. The emphasis on the isolated individual denies the implicit sociality of an activity like driving and this in turn limits the personalisation of experience. Recognition of the social nature of driving allows for greater personal engagement with the journey as well as with others.

The boundaries between individual and social structuring of experience are difficult to draw because they are implicitly interrelated. This will be explored more extensively below through Barbelet's social phenomenology which recognises social engagement in experiences such as boredom. Individuals are active agents influencing and actively engaging with what is socially sanctioned though the structures, institutions and cultural resources available to them are shaped by the priorities of the larger social system.

Barbelet discusses the lack of engagement in events that has been considered a leading characteristic of boredom and the accompanying argument that this is not merely lack of interest but has the added dimension of lacking vital interest and enthusiasm (Barbelet 1999, 634). For Barbelet boredom is not a passive resignation to events but an active discontent, 'a restless and irritable feeling about an absence of interest' that he distinguishes from ennui:

> Whereas ennui may be understood as a feeling arising from a withdrawal from interest, boredom has an additional dimension of agitated feeling that expresses the subject's distress at their finding no interest in their activities or circumstances. (Barbelet, 1999, 635)

A key characteristic of boredom in this account is reflection on the experience. In monotonous activity performed through habit there is typically no agitated restlessness or desire for something to be different, considered characteristic of boredom (Barbelet 1999, 635). In repetitive action as a means to an end there is minimal involvement and no feeling of boredom. It is when repetitive action loses its function or purpose that boredom, involvement of a negative type, arises (Barbelet 1999, 635). Loss of purpose can also arise from overstimulation such as in the 'intensity of metropolitan life'. It is the loss of meaning or purpose that underlies boredom. Not all habitual behaviour leads to boredom then, and boredom involves a state of arousal for Barbelet, however, an 'appropriate interpretive framework' giving meaning and purpose to the action is required for habitual action to be free from boredom (Barbelet 1999, 636). The engagement in habitual activity has a social context in which it is meaningful.

For Csikszentmihalyi (1990) boredom is the result of understimulation through insufficient utilisation of the actor's skills while excessive stimulation, the opposite, can cause anxiety. In his account of flow, engagement is immersion in an activity. Barbalet maintains that different referents are involved in this formulation, which is concerned with anxiety over adequate performance of a task as distinct from anxiety concerning pointlessness in activities. Interestingly the former relates to the comments by men and the latter to those of women in the focus groups indicating the popular social framing of activity in relation to boredom. Boredom for Barbelet also involves a distortion of time-sense so that it is 'precisely the absence of meaning of an activity that promotes consciousness of time as an empty interval' (Barbelet 1999, 637). Work goes slowly while the mind dreams of being elsewhere.

Barbelet's social phenomenological account contrasts with that of Csiksentmihalyi in that there is a sociality that is central to Barbelet's account allowing for boredom as a form of engagement, an active involvement. Sociality is not associated with passivity or the overtaking of the subject by social and cultural demands in his account. Sociality and enculturation are important for the constitution of the subject and social and cultural immersion forms a background to activities, particularly those that are prominent in popular discourse.

Activities such as gambling, risk-taking and even fashion have been popularly regarded as an escape from the boredom of life (Barbelet 1999, 641–42). The idea of an escape is often based on the notion that social control is all pervasive and takes away any spontaneity and individuality: ' ... people value risk-taking because it is the only means they have for achieving self-determination and authenticity' (Lyng 1990, 883). This view is reliant on a popular opposition between individual and society, spontaneity and constraint:

Normal social life is 'unreal' in the sense that most of the individual's action contributes to a social agenda that is little understood and that often appears trivial when examined critically. (Lyng 1990, 881)

The broad social agenda is often alien to the everyday needs of ordinary people who are not able to easily connect with it, however the emphasis on individualism and downplaying of sociality makes social connection something foreign and imposing – the individual is spontaneous and community is constraining.

It is not social control in itself that is problematic or leads to boredom but the form of social control evident in promises and fantasies that cannot be fulfilled. Boredom, while an active engagement with the social context of driving is the result of the expectations fostered by consumption and marketing forces through which individuals are socialised and enculturated. Freedom as escape and extreme thrill and excitement is promoted as the solution to the mundanity of everyday existence cultivating an unquestioning acceptance of the inadequacy of one's own life and the essentially boring nature of everyday existence without the added thrill and excitement that objects of consumption bring. The value of experiences that differ from those that are socially sanctioned is subsequently denied.

The chapter has explored the social significance of boredom and pleasure through the routinised associations of time and place formed in connection with prevailing notions of freedom. Ideas of freedom promoted in advertising help to frame the experience of driving with particular expectations and emotional experiences which will be explored further in later chapters. The purpose of this chapter was to show how boredom and pleasure are implicated in driving cultures and the sociality of driving through prevailing ideas of freedom and further that there are other dimensions of experience that are revealed with closer discussion and examination. Both levels are forms of engagement but the level Morse refers to as distraction is of particular social significance and concerned with a specific type of engagement within the social context of driving, that is, it is a form of agency and a way of being through cars in relation to the social context, and is not merely passive. When the deeper more personal level is considered through the concrete experience of young drivers, varied differences from the socially framed experience begin to appear though it is the social framing that initially dominates discourses of driving. The articulation of roads in relation to ideas of freedom is also a significant aspect of the social and cultural framing of driving as a practice contributing to how cars and driving are approached in particular contexts.

Emotion and desire will be considered in Chapter 4 and social control will be explored further in Chapter 5 where governance will be considered in more detail. The next chapter will consider gender and age as characteristics through which the driver becomes routinised in their relationship to cars and investigate the social framing involved in gendered and age associated driver-car assemblages.

Chapter 3

Cultured Drivers

A significant dimension of driving practice is the way in which the driver is routinised through gender, age, and enthusiasm for driving, in turn leading to a car-person identity. The driver themselves is routinised in the approach to cars evident in the differences between men and women, for example. While cars are made meaningful in themselves and are articulated in particular ways, characteristics of the driver are also important. It can matter what kind of car is being driven as well as where and when it is being driven, but it also matters who is driving. Gender and age are dominant characteristics of drivers that are often focused on by the media and characteristics of personality such as aggressivity and propensity for risk-taking have been investigated in psychological research. In social and cultural terms men and women are expected to drive in particular ways according to their gender. Age is often stereotyped as well, with older drivers being regarded as a safety risk because they cannot 'keep up' with the faster pace of traffic and are considered to lack quick enough responses. Young people are often seen as the rebels of the driving community, tending to take risks, drive too fast and burn the candle at both ends.

Women are allegedly becoming more like men in the way they drive and losing the differences that have made them safer drivers. It is argued here that the way in which men are constructed as drivers is significant in the over-representation of males in every category of road safety, including as pedestrians. 'Combustion masculinity' will be considered the dominant standard of driving where the aggressive and competitive demonstration and exercise of car handling skill is given prominence over caution. Women are increasingly demonstrating this style of driving as it is arguably the dominant style. Comments from young women however suggest that they are maintaining their differences and distinguish themselves from young men's interest in cars. There is a higher fatal crash rate with sports cars than other types of cars and this is attributed to them being driven more aggressively and by younger drivers (Volti 2004, 145). The style of driving that applies to sports cars is the standard of driving for men and the type of masculinity that must be expressed through them. The expression of youthfulness and risk are associated with sports cars as well as prestige and power and it is these that define the standard of driving and are most closely associated with men, particularly young men.

The first part of the chapter will focus on the discussion of gender in focus group comments and the evident expression and dominance of combustion masculinity. Aspects of gender performance such as knowledge and the 'naturalness' of male connection to cars and how these are defended is considered. Gender is considered as a social performance in order to emphasise the social and cultural construction of combustion masculinity and the apparent connection between men and cars. Age is also discussed as another characteristic of the routinisation of the driver.

Finally an analysis of a television anti-speeding advertisement is presented in which there were evident gender differences in attitude. Attitude is considered as a construct that is relevant to cultural examination of driving as a practice in that attitudes are generally shared rather than being purely individual and they indicate a level of contention in the discourses relating to driving. The tension between discourses of road safety and discourses of driving become apparent in differences in attitude and the dominance of particular attitudes.

Dangerous gender performances: 'combustion masculinity'

The chapter discusses data from focus groups on the significance of gender in shaping young people's relationships to cars. Little analysis from a socio-cultural framework has been applied to the specific practices and experiences afforded by cars as they exist within a broader driving culture, especially for younger people. It is important therefore to link larger social categories such as gender with the ways in which gender is expressed through cars. Analysis of focus group discussions showed distinct differences in the ways in which young men and women 'perform' in cars and how cars are a significant aspect of their evolving identities. Employing and reshaping the idea of 'hydraulic masculinities' from the work of Linley Walker, the chapter will outline and extend the concept and reframe it as 'combustion masculinity' in order to highlight the explosive nature of this form of masculinity as it is promoted and framed in social discourse. It will be related to the focus group data indicating the involvement of danger in this form of masculinity as it is expressed through cars. The importance of age and gender as social norms significantly shaping young people's relationship to cars has not been emphasised enough and yet it could make a significant contribution to how gender performance might be appropriately addressed in road safety.

In this chapter gender will be explored in its social and cultural dimensions as an aspect of driving as a cultural practice (Redshaw 2006). Gender is regarded as formed within a context of social performance, and the particular ways in which gender performance is related to cars is considered as a social norm relevant to driving. In theories of driver behaviour (such as Parker, Manstead and Stradling 1995), social norms are considered as informing driver behaviour, however there is a need for greater in-depth exploration of those norms. Drawing on focus group studies the chapter will explore the meanings and shape of gender performance in cars evident in young people's discourses on cars and driving.

Social psychological research in road safety has been based on the theory of planned behaviour with the aim of identifying behavioural characteristics associated with risk and the tendency to commit violations (Parker et al. 1995). In one study it was concluded that 'violations are social phenomena which require explanation in terms of driver attitudes, normative influences and motivational factors.' (Reason, Manstead, Stradling, Parker and Baxter 1991, 65) Social norm is considered to reflect 'the individual's perceptions about what others would want him/her to do' (Parker et al. 1995, 129). The theory reveals important elements of behaviour involved in driving accidents and infringements, however it is limited to an analysis of individual intentions and does not extend to the culture of driving as a whole, and the expectations and ideals that are operating there to influence the individual driver. It does not address behaviours that are implicitly condoned in the social context of driving.

There are likely to be a range of conflicting influences on a young person's behaviour, including parents, other drivers and peers, and broader social forces such as the various media. While social norms are not consciously chosen, they provide 'the horizon and the resource for any sense of choice that we have' (Butler 2004, 33) thus creating possibilities of who we can be. It can be seen in the ways in which young men and young women express themselves in and through cars that young people are actively producing their gendered identities through driving in important senses (Walker, Butland and Connell 2000). At the same time these performances of gender are produced within the context of social influences, particularly those of family and peer groups (Sarkar and Andreas 2004).

The earliest stages of driving are clearly crucial for young people with statistics showing more crashes occurring in the first few months of gaining a licence. The pressures on young people adding to the inexperience already extensively noted in research (Ferguson 2003) include developing the appropriate gender performance. For young men this is particularly acute as their gender performance involves acting in and through cars in ways that are more dangerous and risk related (Harré 2000; Clarke, Ward and Truman 2005). Young men have also demonstrated a greater self-enhancement bias whereby they consider themselves better drivers than their peers (Harré, Forrest and O'Neill 2005; Gregersen 1996), and have been found to have a less positive attitude to traffic safety and rules than women (Laapotti, Keskinnen and Rajalin 2003).

It is evidently an important part of the social performance of young men to demonstrate a willingness to take risks in cars and this is an additional pressure that needs to be carefully addressed in road safety. The idea that men are more likely to be able to control a car is a familiar claim even though men have more crashes and more serious crashes (Clarke, Ward and Truman 2005). There are considered to be differences in the types of driving skills males and females focus on (Laapotti, Keskinnen, Hataaka and Katila 2001) but this is not so much an issue of competence and incompetence, as it is usually framed, but of different relationships to the car and the surrounding system of traffic.

Social researchers have drawn attention to the need to consider the role of gender in road safety for some time. Linley Walker in particular has outlined forms of masculinity associated with cars. Walker, Butland and Connell (2000, 158) warn against stereotyping boys as being a certain way and girls another, drawing attention to the 'diversity of masculinities'. It is important to stress then, that forms of masculinity are social and cultural and produce norms of behaviour that are not biological or fixed. As Malcolm Vick points out however, even though there are a range of masculinities in Australian culture, those that are most often emphasised are those that 'place a high value on risk-taking, bravado, skilled performance with machines, rule-breaking and other forms of challenging authority and convention' (2003, 35).

Walker's research focused on particular types of masculinity amongst juvenile offenders in Western Sydney (Walker 1998 and 1999). She was concerned with car culture as a form of 'protest masculinity' amongst young men who were marginalised in the labour market, deprived of material resources and failed educationally by society as a whole (1999, 178). 'Hydraulic masculinity' refers to the investment of sexuality, as a 'naturally gushing force with an uncontrollable and addictive power in men's lives' in cars, which provide a medium for male admiration and the expression of competitiveness, performance, power, control, technique/skill and aggression (1999, 183). In this account, masculinity is structured in relation to cars which play a significant part in the formation of identities and the expression of aggression and competitiveness, an association that is socially constructed and popularly reinforced, but nevertheless excused as 'natural'.

In this chapter it is proposed that the 'hydraulic masculinity' outlined above be referred to as 'combustion masculinity' since the key is the explosive nature of the apparent drive to express themselves through cars. While hydraulics involve the movement of liquids and the use of liquids under pressure to provide greater force this involves a more controlled expression of masculinity as will be explored in a later chapter. Combustion or 'fire power', as it is referred to by Virilio (1986), is explosive and is associated with violent excitement, agitation or discontent. It

Figure 3.1   Men gush over the Toyota sports concept car unlikely ever to go into production

requires ignition or rapid oxidation and the constant burning of fuel. Virilio refers to combustion powered speed as the 'binomial fire movement of implosion and explosion', the power to penetrate and the power to destroy (1986, 133). Combustion power is noisy, invasive and a dominating force capable of great destruction and it is both the source of power and an expression of power.

Combustion masculinity is the standard by which young men are expected to approach cars and by which their early experiences of driving are informed. Rather than being related to a particular class of young men, the pressure applied to many young men to indulge in risk-taking behaviours of varying degrees, crosses class and cultural boundaries, and pertains to the ways in which cars and driving are given meaning through advertising and other forms of media. Comments from focus groups illustrate the extent to which the identification of driving skill with masculinity has encouraged a particular approach of young males to driving that puts them at greater risk of injury and crashing than the approach of young women.

Gender was a major theme in the focus group study drawn on here and responses to all questions were analysed according to gender differences in meaning. Focus group questions were open-ended and asked about access to cars, the importance of cars, how they are used in the young people's social lives, knowledge of cars, enjoyment of driving and characteristics of cars and how they should be driven. Discussions showed distinct differences between young men and women in how these experiences were described as well as how they were lived. The following discussion draws primarily on how the young men and women portrayed their driving and their relationship to cars.

The analysis of focus group discussions showed distinct differences in the ways young men and women talked about cars and driving and a tendency on the part of the men to deny that women have anything like the experiences with cars that they do, as we will see in the following. This chapter will draw on comments and discussion that illustrate the way in which young men expressed their superiority as drivers and how that could lead to a higher likelihood of young men engaging in dangerous driving. In all focus groups there were young men who expressed a superior relationship to cars that they regarded as a need and as 'natural' to them as males.

Not all young men in the focus groups however, expressed this superior relationship or implicit connection to cars and driving. A few men even commented that they did not like driving and avoided it and a few were quite reflective and insightful about their earlier driving experiences. The main themes to be discussed here are the perceived connection between men and cars as natural, differences between men and women in knowledge of cars, driving performance and young men's reflections on early experiences as drivers. The ways in which young men defend and assert their superiority as drivers is of primary interest in each of these themes.

A 'natural' connection

Men have long felt a stronger connection to cars and the experience of driving, and express that connection as more authentic and 'natural', as an experience of man-

and-machine. The men in the Shellharbour group described the freedom of being alone in the car and how that allowed them to 'hoon around' in a way that felt 'natural':

> Man 2: I love driving. I love ... like, when you're driving to and from work, it's usually only you in the car. And you can just do whatever you want! And you just cruise along and you can be as much of a hoon (as you want) or have a nice cruise, whatever you feel like.
>
> Man 1: You're in your element.
>
> Man 2: You just go ... yeah, it's natural.

A number of young women in the focus groups seemed keenly aware of being excluded from an authentic experience of driving because they were women. One young woman in the Bankstown group recounted how she had learned to fix her car because it kept braking down in the traffic. She even referred to being able to 'treat' the problem when her car overheated. Immediately following this story a man in the group commented, 'The difference sometimes with guys and girls: see, girls don't want to give it a chance. They don't want to get all dirty.' Another woman in the group retorted, 'I love getting dirty!' The man nevertheless continued on the same tack:

> But with guys, if my car breaks down, I'll try to fix it before I take it to someone. If I can't fix it, then I'll take it to someone else to fix it. But girls sometimes, they don't want to take that chance.

The first young woman who told the original story came back with the point that many men, such as 'businessmen with their suits and their computers and clean fingernails' would also not want to 'get their hands dirty'. In this exchange the young man, as with men in other groups, seemed to want to invalidate the claim of the women to any authentic car experience exemplified by 'getting your hands dirty', purely on the basis of gender and in the face of evidence to the contrary, in the form of accounts of experiences from women in the groups.

Many of the young people were aware of the difference between what were regarded as 'guy's' cars and 'chick's' cars. The gendering of cars was evident in statements from women as well as men in the groups, though this was troubling for some. One young woman explained that the car she wanted was considered a 'butch' car. This was partly related to the power of the car whereas colour, she pointed out, could be indicative in other ways:

> My dream car is a Ford GT. My boyfriend said, 'That's not a girl's car! You can't! It's too butch for you.' I said, 'No, I can drive it, if I could drive that, I would drive that!' But he's like, 'It's not a girl's car.' What's a girl's car? Pink? (laughter) When anyone sees my car they go, 'Oh, that's such a chick's car,' because of the colour. (Woman, Bankstown, 20 years)

Figure 3.2   Holden Astra preferred car of young women

Driving 'like a hoon' was defended as a right that young men are entitled to. A younger man from the Blacktown group stated the need for the correct image:

> Yeah, being a male it's a big thing to have a decent car. I always before I got my car, I was always pretty embarrassed saying, 'I'm driving Mum's car.' It's a big thing. Not necessarily status but just kind of who you are. If you're driving a little car … like me and my mates were very sarcastic about it. We kind of looked at them as though they're lower on the food chain, shall we say. It also does something for your ego, too. You say, 'My car can do this, my car can do that,' you feel bigger and better than everyone else.

This young man was well aware of the competitiveness involved in the kind of car a young man has and how he drives it. The connection with 'who you are' demonstrates on the one hand, that the car is related to identity in a significant way, and on the other that this connection is claimed as 'natural'. The right car for the right image relates to social esteem and the importance of adopting the image of masculinity that others appreciate, but needs to be seen as a connection that is normal and innate.

Another young man related the memory and significance of his first car being small and powerless:

> … I seriously attribute it to me still being alive, because I know when I was 17 I was an absolute hoon and I wanted to get myself a VK Commodore, and I had to have a manual car because, you know, I can drive a manual so why drive an automatic … and I know I

would have wrapped myself around a tree because the speedo says you can do 220 ks an hour, so I want to see if it can do 200ks an hour! (Shellharbour, 24 years)

The men in the focus groups were more likely to talk about having a 'natural' and more authentic connection with cars and the experience of driving than were the women. Driving 'like a hoon' was associated explicitly with men in all the focus groups and was often regarded as a 'natural' way for young men to behave because of their implicit connection to cars. Young women talked about 'having time to themselves' and the independence of having a car. They did not talk about driving as 'you feel like' or doing 'whatever you want' with a car.

Knowledge of cars

Knowledge is central to any convincing performance of social or technical competence and it was generally assumed in the groups that men have superior knowledge and mastery when it comes to cars. The young women in the focus groups sounded quite knowledgeable about cars at times and were keen to show their knowledge, yet the kinds of knowledge young men and women were concerned with were often different. Knowledge could mean knowing 'how fast a car is' (Male, Warrawong) or 'how to control it' (Female, Warrawong). Young women in the Bankstown and Blacktown groups talked about needing to know the capacity of their cars because they 'thrashed' them. For most of the women, however, knowledge meant knowing some of the basic things about cars in case anything goes wrong and knowing how to deal with unfamiliar cars. For the men it was more about knowledge of the power and ultimate performance of the car and how that could feel and also how it reflected on them.

Figure 3.3   Young men under the bonnet in the car park of a hotel

One young woman said she had been given a crash course by one of her friends and knew how to change a wheel, oil and indicator lights (Bankstown). She said that if she did have a problem she would have some idea of what it would be. Others in the group said their fathers had taught them to change tyres, the oil and fill the water. Knowing cars also meant knowing whether the car had power steering and how powerful it was:

> But I think it's good to know what cars you can drive and what cars you feel comfortable driving, rather than, 'Oh yeah, it looks good, I'm going to drive it.' To kind of know which ones you're capable of driving. (Bankstown)

The women talked about 'flattening' the accelerator and what happened in different cars, and knowing the difference between different types of cars. Most of them acknowledged that having some knowledge was desirable, especially if they had an older car, but that they would not necessarily use the knowledge. A few said they would just call the AAA (NRMA in New South Wales) or a family member or friend. One woman stated:

> I'm happy, like I said, with the basics, but when it comes to the technical stuff I'm not really fussed. I don't want to get my hands dirty in the bonnet. I'll open it, but that's about it. I'm not really into that sort of thing. I'll call my dad or take it to a proper service person, a mechanic. (Bankstown)

Women in the Goulburn 1 group talked about knowing 'how the water squirters work' and how to put air in the tyres, and also expressed a lack of interest in touching anything under the bonnet. As one woman said, 'I know where it is … like, Dad comes out and gives me a quiz, where's this, where's that. And I'm like, "Yeah, there …"'. Some of the males were equally uninterested in knowing about how cars work or dealing with problems, but it was a more typical female response. One young woman demonstrated the extent and focus of her knowledge of cars:

> So you have the two different sides of the blinkers (laughter), and you turn on the windscreen wiper! (laughs) That's annoying, too, so you need to kind of think … even my mum, who's been driving for ever, she still gets confused because the X-Trail and the Astra are on different sides. (Bankstown, 19 years)

Young men were highly unlikely to talk about where indicators and windscreen wipers are located and were more likely to demonstrate knowledge about performance and skill:

> We talk about top-end and acceleration and stuff like that, and they get off the mark quickly, but we'll smash them once we catch up to them, and stuff like that. Just getting it sideways every now and then, taking off really hard, … (Male, Blacktown, 18 years)

By maintaining that they had superior knowledge the men were validating their claim to have a closer relationship to cars that was more authentic than the women and this in turn validated the style of driving most associated with men – speeding, 'being a hoon', demonstrating skill and competing aggressively with others. Combustion masculinity is being defended as a necessity. This was clear in the defence of driving performance where men maintained their superiority.

Women's responses to spending time with young men who went to places where cars and the modifications to them are displayed such as The Rocks in Sydney, were indicative of the level of interest most women maintained. They could not understand what all the fuss was about and while they might be interested enough for a time in being there with a boyfriend their interest soon waned:

> It's really overrated, because once you start to know what they're talking about you just go, 'Oh, is that all? It sounds really boring to me!' Why did I take all this time to learn and ask so many questions and all they're talking about is the inside? (Female 3, Lawson)

> It's really boring driving down George Street, bumper-to-bumper just to go park your car and lift your hood! (laughter) [A friend] used to do it with one of her friends and she found it absolutely boring! (Female 3, Bankstown)

> Like, I find that very boring. But they do that, just compare cars and pick up chicks. That's what they all said to me. And they all think the cops are racist down there, as well. ('My car's just laying there! It's this high!') (laughter) (Female 4, Bankstown)

One man talked about how he related to the modified car scene as a way of spending time with other men and learning about cars:

> Yeah. I was like a newbie, a person just starting to get into modified cars, but all my other friends had been in it for longer and had modified cars. They know a lot more, so I'd just listen, pretty much. It's more like hanging out, you know? Something to do when you're bored. (Male 1, Fairfield)

Some of the women were interested in knowing a bit more about their cars than others but most were simply not interested in the 'insides' of cars and how they could be optimised whereas for the young men car performance and how to maximise it were like a rite of passage.

Driving performance

While performance is an often discussed feature of cars, it is not only the performance of the car that matters. The power of the car was significant for many young men as noted above, but this often had to be accompanied by a driving performance that matched the vehicle. Cars appeared to be significant for some of the young men in that they suggested important things about themselves as males, and this could also involve demonstrating a style of driving that is considered more fitting for young men.

The contrast in the way the young men and young women spoke about their driving performance is evident in these comments about dealing with slipping tyres:

> The other day I was turning off Clinton Street into Sloane Street, it was just raining a little bit and I was going round a corner and the tyres started to slip and I thought, stuff

it! And I gunned it! Woooh! It was just sittin' on steel, and there were all these cars starting to bank up behind me, and I was just going, yay!! The tyres were going and I wasn't moving. I wasn't going sideways, I wasn't going forward. I was just sitting there … (Male, Goulburn 1, 18 years)

In this account the young man expressed excitement and sheer pleasure in having to deal with a situation and a vehicle that was difficult to control. He described the wheels just spinning on the spot to the point that it felt like the wheel hubs were grinding on the road as the vehicle failed to move forward. The experience for this young man is bodily and contrasts with the careful consideration of the young woman in the following:

> I think knowing your car, as in exactly how it is. Knowing the grip of the tyres and things like that. I've got a few cars in my family and I just drive around whenever. And going by the different cars, you can tell when you take a corner. One flies out but the other one doesn't. So, just knowing the car, pretty much. And the roads. That's very important, because if you're in a place where you don't know the roads and you're speeding, there's no chance. You're gone, because if there's a blind corner, you're gone. That's how I feel. Know the roads, know your car, you should be right. (Female, Warrawong, 19 years)

For the young man the experience was exciting, he 'takes it as it comes', and enters into the moment with the car and the road. He remained in control in his merging with the vehicle, whereas for the young woman the over-riding sentiment was caution, getting to know the situation, the car and the road in a very deliberate sense. It is not just the way each speaks about their experiences that is of interest, it is also the experience itself and what each was inclined to do that was different. The young man 'gunned it', while the young woman preferred to exercise forethought and vigilance. The contrast is perhaps between enjoying the sensation of 'sittin' on steel' and knowing the car and the streets in a reflective sense.

A number of the young men wanted to make it known that they were better drivers. In all of the focus groups there were discussions where the relationship of men and women to cars was distinguished and clarified. Challenges to the superior driving and relationship to cars on the part of the men were taken up and dealt with convincingly enough for many of the young women to eventually agree with them. These challenging incidents were mostly initiated by women, whereas men tended to focus on maintaining the gender distinction. Women related stories about their involvement with cars that appeared to challenge the distinction between men and women in relation to cars and driving, but eventually the gender distinction was maintained.

When a young woman in the Warrawong group exclaimed that female drivers were safer, one young man responded; 'Not the ones I know mate! I don't really like going in cars with them.' When asked why, he said at first he did not know and then referred to 'their' braking strategies: 'they just brake at the last moment and go over the gutter.' A woman in the group countered by arguing, 'statistics say that males are the biggest problem.' The same young man responded that males are 'obviously stupid when they are in a car' to which a woman declared 'there you go'. The young man went on to outline the difference between men and women drivers in some detail:

> Yeah, but I'm just saying that women don't know how to drive! Like, if I'm in a car and I'm speeding – it's stupid if I'm speeding, but say something happens, I can maybe control it, you get me? A woman, a girl, something that happens to her. I know a friend, a chick, she lost control and she smacked it into a pole. And one of my mates was speeding. He lost it and, like, he controlled the car. I'm not saying that women are worse drivers. How do you say it – a man knows cars better than a woman does. That's just what I'm trying to say.

In this statement the man who 'loses it' is regarded as remaining in control whereas a woman is not because 'a man knows cars better than a woman does'. At the same time he said he was 'not saying that women are worse drivers'. He was trying to find a way to express the 'obvious' gender differences in how men and women express themselves in relation to cars. In his example the woman crashed into a pole while the man only 'lost it'; even though he was speeding, he was able to regain control. The fact that men were 'stupid' in cars was not a problem for the men in the group, nor was the fact that they tended to speed or lose control. Somehow the crashes and 'stupidity' men were involved in were excused because men were seen as having a stronger affinity with cars. This affinity was part of how they engaged with cars and was demonstrated in their driving style, and this was regarded as part of being male.

The young woman who initially challenged the young man in this group then agreed with him stating; 'You could have a point there.' Neither she nor any other women or men in the group took him up on his statement that 'females just don't know how to drive'. He convincingly redefined driving to suit male styles and was encouraged to go on:

> You see a woman there, she's got a flat tyre and she's just standing there at the side of the road! (laughter) And she waits for someone to help her. A male will know what to do.

The young women in the group agreed and he continued, defining male capacities:

> I'm just saying we can react quicker. We know what we can do. A woman will just let go and hope for the best, and accelerate or brake.

He was not quite saying that women were incompetent but his comparison suggested that the difference between men and women was their skill and this stemmed from men's implicit and more intense engagement with cars. The gender distinction was kept in tact in this discussion with the men maintaining superiority over women in their driving. The male view that men 'react quicker' and 'know what to do' while women 'hope for the best' was not challenged even though the men's style of driving might be more dangerous because it involved speeding. A dangerous style of driving was implicitly equated with better driving.

Reflections on early driving desires

A few of the men reflected on their earlier driving and how that was crucial to their sense of themselves with one recognising the potential consequences. He related his desire for the kind of car that would adequately express combustion masculinity:

> I know when I was 17 I was an absolute hoon and I wanted to get myself a VK Commodore, and I had to have a manual car because, you know, I can drive a manual so why drive an automatic ... and I know I would have wrapped myself around a tree because the speedo says you can do 220 k an hour, so I want to see if it can do 200 k an hour! (Shellharbour, 24 years)

A powerful manual car would have allowed him to demonstrate his skill and mastery as a male and would also have facilitated the expression of the 'natural' gushing power associated with masculinity. The importance of identifying as a young male with a car like the Commodore was recognised by this young man who later became aware of the enticing danger it presented. The age of seventeen was clearly a crucial stage for him and he appreciated the fact that he was not able to indulge in the desire for a powerful car at the time.

Another 23-year-old man from the Fairfield group referred to the 'old days' when he was younger and more into cars:

Figure 3.4   VL Commodore a model popular with young men as a first car in Australia. The cars are often modified to the tastes and enthusiasm of newly licenced drivers

So you just get a normal, decent car, but back in the old days I used to like knowing how your car performs and all the dynamics and that sort of stuff, but these days I'm starting to forget everything because there's just no time to keep up with it.

A number of male focus group participants stated that they were 'into all that modified car stuff' when they first got cars but now that they were older they were more interested in comfort and convenience. The modified cars many of the younger men are interested in take quite a bit of effort to keep on the road in their characteristic uniqueness so that they are eventually less inclined to make the effort. This could also have some effect on the way they drove:

I do drive like an idiot sometimes, but I think I've settled down a lot now. When I first got my P's I was just an idiot. Now when people are in the car I think I drive a lot better, but when I'm by myself I still give it a little bit just to feel the G's or something. (laughter) I don't know. When people are in the car, now, I drive a lot safer than what I used to, and I've had my licence for two years now. (Male, Goulburn 1, 19 years)

There was still a need for the young man to express himself as a male in the car when he was alone but he was now prepared to be more conscious of passengers in the car at least.

The young man from Shellharbour went on to talk about the first car he did end up with, and its lack of power:

And we've been out and looking a bit, me and Dad, to try and find a car for me, and we went to the auctions, and the week before they had a whole heap of VK and VL Commodores, and they were going for under three grand. So, it was like, 'Right. If we go there, maybe we'll get lucky this week.' They didn't have any there but they had the Bluebird there, and it was something completely different and it was just very basic. It had 22 horsepower, and if you had a tailwind and you were going down a hill, I think that had got it up to, like, 75 miles an hour, which is about 120. So it was one of them cars that you really couldn't drive fast, and it just suited my needs at the time.

The car was lacking in power but suited his needs to get himself where he needed to go and he clearly appreciated this at 23 years of age. His friends nevertheless found a way to code the car he had as male by invoking animal power:

It was just it was my car and all my mates saw it and they were like … instead of calling it the piece of crap that it probably was, they were like, 'Yeah, that's Dazza's beast!' It was something completely different. (Shellharbour)

In this case the masculinity of this young man could say something about the car even though it was not the most masculine car or the kind of car that was regarded as typically male, like the Commodore. It became a 'male' car in being claimed as such by his mates who thereby made it socially acceptable. The Bluebird could not respond to the desire for gushing power with its limited speed and power. Commodores are also limited but their limits tend to be well beyond what is acceptable and manageable on the roads, and this is part of their appeal.

The young and the old

Differences in age were also evident, even between 17 to 18 year olds and those aged in their early twenties. As previously noted a young man from the Fairfield group who was 23 years old refered to the 'old days' when he was younger and more into cars:

> ... back in the old days I used to like knowing how your car performs and all the dynamics and that sort of stuff, but these days I'm starting to forget everything because there's just no time to keep up with it.

He stated that he had forgotten and there was 'no time to keep up', rather than that he had lost interest, which could bring his virility into question. A few men in the focus groups said that they were 'into all that modified car stuff' when they first got cars but when they got older they were more interested in comfort and convenience. There were young men who were the exception and had carried on the interest in cars into the late twenties.

Age is thus an important influence on the type of car driven and how the driver feels when driving it, according to the young people. The following comment relates explicitly to the connection between type of car and style of driving where age is a recognised aspect:

> I think the sort of car you have can reflect what sort of driver you are, as well, especially with young drivers. You can tell if a young man or a young woman is driving the car by what car it is, some of the time, and you expect to know what sort of driver they are, to a point I think, as well. (Female, Blacktown, 19 years).

Young people often negatively stereotyped 'old' drivers claiming that they drove too slowly, could not keep up with the traffic, and did not pay enough attention to what they were doing:

> I associate the old EH Holden with bad drivers, because there's usually old people in them and they don't put their blinkers on and stuff. (Male, Goulburn 2)

Age is an important factor in the stereotyping of drivers and contributes to a debate about who has more right to access the roads. Arguments arise over the suitability of older people having licences when they 'can't keep up with the pace', lose flexibility and therefore response times are slower and they are not able to negotiate the traffic as well requiring others to allow for them. Research has suggested however that older drivers are a calming influence on the roads, do not have more accidents but are more easily injured and that when they are forbidden from driving they are more vulnerable to being injured crossing roads and accessing public transport (Hakamies-Blomqvist 2003, Langford, Methorst and Hakamies-Blomqvist 2006). Mobility issues at all ages are important social issues and require consideration beyond the car but must also take into account the quality of life that adequate mobility brings. Too often mobility issues are decided based on the standards demanded by the most able bodied and aggressive road users who have access to vehicles, primarily men.

Cultural costs

The cars and driving performance that are considered appropriate for young men could be extremely costly for them. It is not just the power of the car that is significant, it is the ways in which the power must be demonstrated and mastered amongst young men. Not only is risky driving considered acceptable amongst males, having a male passenger in the car makes it more likely that the driver will take risks (Simons-Morton, Lerner and Singer, 2005; Ulleburg, 2004; Harré, Field and Kirkwood, 1996). It is not only knowledge of cars and their willingness to get their hands dirty that makes men different in their relationship to cars. It is also the association of particular masculinity with cars and that it has to be expressed in particular forms. The investment of masculinity in cars is strongly defended as 'natural' and efforts to exclude women were evident in order to maintain the apparently more 'natural' association of men and cars.

Combustion masculinity, it has been argued here, is strongly age related though it appears in other age groups and women may also demonstrate it. A large number of young men feel drawn to express combustion masculinity when they first have access to cars. For many this later changes, but combustion masculinity remains the standard for the expression of masculinity through cars and to some extent for driving generally, fuelled by car advertising which constantly employs images of combustion masculinity. It could be said to be the dominant social norm of driving informed by other social norms relating to gender and gender relations.

It is important then that masculinity in the forms related to cars is regarded as socially constructed rather than as a fixed biological characteristic of men. In order to change the way young men take to cars, the forms of masculine expression through cars need to be addressed and considered. It is not popular to associate maleness, cars and caution but it could become more acceptable if advertising is approached differently and there is greater acknowledgement of the constructions of masculinity and the associations with cars and driving.

Taking some of the emphasis related to 'good driving' from skill in handling corners fast and putting more emphasis on caution and regard for the social environment inside and outside the car could help to change the desires expressed through cars by young men. The desires that are encouraged socially are the free expression of youthful exuberance and demonstration of extreme handling. This needs to be modified to emphasise the planning and thought needed in driving as well as concession to other road users and impact on the social environment. Social and cultural influences need to be recognised and addressed and not merely treated as innate drives.

Young men and women were both keenly aware of the demands to conform to gender appropriations in terms of car type. Age differences were also considered important. While cars offer the pleasure and freedom of movement, then, they are also implicated in shaping and restraining those engaged with them. In many of these accounts it is apparent that there is an evident coerciveness in the way young people feel they need to relate to cars. This was not necessarily troubling to many of the young people, but it clearly has consequences that are evident in the higher representation of young people, especially men, in crash statistics. The impact of dominant social norms such as combustion masculinity in the shaping of the practice of driving needs to be acknowledged and confronted. It cannot be assumed that the dominant norms operating for young men are the ones we might wish them to have – those conforming to the rules of the road, caution and the needs of others. The social norms through which their behaviour is guided and given meaning embody attitudes that are inimical to the norms we expect to be dominant as will be explored in the following discussion of young people's responses to a television safety campaign advertisement.

Embodying attitudes

Cars and social norms such as the individualism promoted through them, are central to consumer cultures. Driving cultures, related to particular kinds of cars and drivers and the driving styles that go with them, express broader social and cultural embodiments, attitudes, expectations and beliefs. Attitudes can be seen as informing and encompassing cultures and the social norms through which they are embodied. They can produce and express habits of practice but are also often about something controversial and unsettled or ambiguous (Bilig 1987). They are not so much 'below the surface' as part of the surface that is everyday and unseen and yet also defended and justified. Attitudes are not normally the sort of construct that cultural researchers are likely to employ however they can be useful in denoting certain orientations to the world and specific objects and relations within it.

The important point here is that attitudes are not merely individual but are derived from a social and cultural context. It would be rare for a person to maintain an attitude that is not found amongst others in some form. This does not mean that individuals are just made up of social and cultural constructs. Individuals choose attitudes in association with their values usually having some connection to family, peers, profession, locality or special interests, for example. As Judith Butler notes however, we do not fully choose our values and social norms, rather these form the horizon and resource for our choices (Butler 2004, 33).

Attitudes, it is proposed here, could be useful in highlighting contradictions between sets of values and social rules. Rather than being permanent and unchanging they can be situational and specific, related to particular practices and objects and expressing dispositions at odds with the dominant ideals. In the broader social and cultural context within which young people begin to engage with motor vehicles, it is considered desirable for young males to feel at risk in order to make the most of their experiences, leaving little margin for error (Harré 2000). The ostensible standard for driving on the roads is safety and caution and obedience to the rules. There are other standards operating however, that conflict with this and favour demonstrations of skill and handling, high speed and even flirting with danger as we have seen in considering combustion masculinity as a social norm.

In the following, discussion of a television campaign advertisement from a number of focus groups will be presented showing different attitudes to the issue of speed. These attitudes are orientations that are influenced by gender as well as location (city and country) and views about the capacities of drivers and cars.

Attitudes could reinforce or have an ambiguous relationship with social norms as can be seen in some cases in the following analysis of responses to a television anti-speed campaign.

The Police 'Reverse' advertisement was a speed campaign ad in which a family is shown in a car in the process of crashing. The crash is shown in some detail and in reverse. As the car comes back to driving along the road it is shown being pulled over by a policeman. The message is that it is better to be fined for speeding than to crash. The driver is male, late twenties to early thirties, and his wife, also late twenties to early thirties, is sitting in the passenger seat and there are two children in the back seat. The parents could be considered to be middle income paid professional/s (teacher or other government employee). The slow crash scene starts with the wife being shown hanging in her seatbelt apparently dead. The male driver is gasping at the sight of his wife as the car swirls back into the roll across the road. The car rolls in reverse back to being upright only this time it is stopped next to the road. There is a police officer standing beside the car and the driver is being booked. While he is being booked the driver says to his wife 'Sorry, it's the car, it just wants to ... [go]'. He gestures with his hand indicating a forward momentum.

Concern for others and 'I know what I am doing'

A number of views were expressed in the focus group discussions about this advertisement, however there are both commonalities and strong contrasts in the responses of men and women. Many of the female participants in the focus groups identified with the woman in the passenger seat in the car, and commented on the male driver who suggests that he 'couldn't help speeding, it was just the car'. The women were more likely to relate to and express a concern for the welfare of others. Men were more likely to regard the issue of speed as over emphasised and to trivialise it by broadening the message to 'don't crash', or to regard the problem as boredom, distraction or tiredness, and not speed.

The men were generally incredulous that anyone could crash the car on a straight stretch of road. In the Bathurst 1 group some males decided that the driver must have been speeding at an extreme level and considered the message of the ad to be 'don't crash'. They then decided that the problem was inattention, not speeding:

> M2:   How could this guy ... rolling his car on a straight country road with a good surface on it, so you just think, like ... you'd have to be not just speeding but, like, strapping on the afterburners to lose control of the car. So, I'm not entirely sure what the ad tells you what not to do! Don't drive like a moron?

Here the 'moron' is someone who cannot control the car at speed. The subtext is that men who know how to drive would not lose control even if they were speeding. The attitude here is that speed is not a problem in itself but a matter of the control capacities of the driver. Another man then took this argument further by suggesting that inattention was the issue:

> M3:   Don't crash your car! Doesn't that sort of show that it's inattention, not speeding, that's causing the accidents?

A few kilometres over the limit is not regarded as significant. Distraction is seen as the problem and speeding is not considered as contributing to the crash. The point that distraction is always possible, while speed can be controlled to reduce the chances of distraction having such drastic consequences was not discussed.

Young women in another group also considered the implications of the male attitude to speed while the males accepted and agreed with the parody, appearing to include themselves as being able to 'drive good':

> F1:   Those guys go, 'I'm invincible, I can drive good. I'm not going to crash just because I'm going fast.'
>
> M2:   [It's saying] slow down on open roads. Country roads.
>
> F1:   There's nothing that you can see that's actually caused it, so it's obviously just the speed that's made it happen. So, it's kind of: safe driving, do the speed limit. (Mountains)

This young woman had no problem accepting that speed could be the cause of the crash even while the men were commenting that it was the type of car that was a problem and agreeing that they can 'drive good'. The men made no comment on the statements of the women and brought up fatigue and boredom as the issue saying 'it's open road', 'he doesn't have to think or anything' so that the driver is not concentrating. They concluded that the message was, 'A car can crash anywhere'. This leaves in tact the position that it is the driver in control and his level of skill that matters and it is skillful manoeuvring that avoids crashes.

Slowing down on open roads was replaced with the idea that open roads do not require you to 'think or anything', to reinforce the argument that fatigue was the issue, and lack of concentration, not speed, at least for the skillful driver. The message was reduced to the idea that a car can crash anywhere, a very broad and particularly benign message, especially since this was not related back to the potential for speeding to add to the possibility of a crash.

In other groups the males also maintained that the issue was not speed but tiredness or boredom. While it was accepted that you could crash on a straight stretch of road if a kangaroo suddenly appeared on the road ahead, the discussion came back to the idea of boredom. According to one man there is nothing to look at and you are 'scattered', 'just driving' (Fairfield). Speed was considered as a way of dealing with boredom rather than an additional risk factor.

In the Bathurst 2 group one male initially made light of the dramatisation saying 'She looks pretty good for someone who's been in an accident ...', while another stated the message of the ad:

> You've got two choices when you speed. You end up with a fine, or you end up dead. That's the impression.

The women meanwhile found the message 'kind of scary when you actually think about it and think that it could happen to you'. They noted that 'the guy just wants to go "it's just the car"'. The men related to having a 'foot full of lead' suggesting both that the car influences the driver and that the driver has a propensity for speed which he is unable to control. A woman immediately picked up on the men's comments in the ad saying 'It's so typical of males' and one of the men defended the driver's position:

> M1:   But on that sort of road, you wouldn't do something like that. It looked like a big, open road.

> F1:   Like, it's just males! They blame the car. They won't take the blame for it! (laughter)

> M1:   Yeah, you get tempted. If it's a big, long, straight road out in the middle of nowhere, I would probably be going faster than the speed limit. That's just the way I drive, I suppose.

> F2:   I do ten percent over the speed limit because I don't think I'll get caught (laughter) and then I add a few more! (laughter)

Boredom and being 'tempted' was again identified by the males as the problem, rather than speed. The women's comments did not appear to make any impact on the views of the men. The men suggested that a long straight stretch of road was precisely where it makes sense to speed and you would be unlikely to crash. The long straight stretches of road made the speed feel trivial, it feels like 'you are not getting anywhere' and 'you are still seeing the same thing'. The women eventually agreed with the sentiment of the males.

Speed for these young people, who do a lot of driving in the country, was part of dealing with long drives. The consequences of speed were not easily drawn out. They all admitted to speeding, although the women noted the view of the men that they were not responsible for their speeding. This suggests that the women were more likely to see the consequences of speeding and the effect on other people than the men, who continued to maintain that it was necessary to speed in order to relieve boredom and that it was not a safety issue.

It was clear to the women that the male driver was putting the responsibility on to the car, and that most men would suggest that they would not lose control in such a situation, even if they were speeding. Women in all groups expressed a similar view regarding the comments of the male in the ad.

Speed was clearly related by the women to recklessness and lack of regard for the safety of others, which was seen as typical of men: 'It wasn't me, it was the car! It's nothing to do with me!' (Redfern 1) They do see the ad as suggesting that there could be implications of speeding for others. 'It's just showing that you could be that driver one day. You could put your own life as well as others at risk for you to speed and be reckless'. They thought that it was about more than 'getting caught by the police', that it also encouraged you to 'look at the big picture' and consider that there might be other consequences.

The young women in this discussion focused on the welfare of others and connected it to the idea that any driver could be in that position. They then went on to discuss how they would feel if they killed someone in the car with them, and the experiences of friends. They stressed the point that speed may not only have an effect on the speeding driver, it may also effect others. One male responded that he also considered others:

M1: That's what I do when I drive. I'm confident I'm a very good driver. It's the other people I worry about. I worry about other people's driving. Because when I drive, I look, you know? Because some people, like, they don't really look and stuff. So I always take steps. (Redfern 1)

He expressed the prevalent male attitude that the women are referring to in that it is the driving of others he worried about which allowed him to shift the potential danger to others. He suggested that he did take responsibility for the speed he was doing and did consider others but this was by way of reassuring himself of his own confidence and being aware of what other drivers were doing. He was watching out for his own safety by being aware of what others were doing. In this he was considering others as a threat that he has to watch out for while he did not appear to regard his own driving as likely to be a threat to others. A woman agreed saying 'you have to drive for others' in case they do something unpredictable and another woman then extended that to the idea that she was especially careful not to be reckless with others in the car. This woman then pointed out that 'driving for others' should also include concern for people in other cars, that your own driving may have consequences for others.

One man from the second Redfern group showed some reflection on his own thinking, noting that he was likely to think it would not happen to him, even though he recognised the speed as the driver's responsibility. A young woman was then encouraged to share her husband's attitude:

M1: I suppose I just don't think it will happen to me. I mean the car only goes as fast as you let it.

F2: Oh, I can relate to that in terms of my husband, he's like that. It's like, 'Well, you've got to go with the flow of the traffic!' And it's like, 'No, you go with the speed limit!' So it's kind of like, 'The car just wants to go faster!' Oh, right. Yeah. OK! The car's got a personality! (laughs)

While young women brought out the issue of the potential to harm others and related stories of the effects of car crashes on friends and family, none of the young men related to the advertisement in this way. They were more likely to suggest that speed was not the real problem, that their driving was good enough to avoid crashing if speeding, and it was other drivers that they needed to be wary of. They considered speed and the message of the advertisement in more general and abstract terms than the young women who were more inclined to consider it in very concrete interpersonal terms.

The young women were able to critique the view of the males that they were 'good drivers' and therefore unlikely to crash, even if speeding, though they were not able to confront the discrepancy in the argument of the males in the focus groups – that speed adds to the potential of a crash when you are tired, bored or distracted.

Both males and females related a tendency to speed that remained unchallenged by the advertisement even when the message was recognised and they could relate to it.

Speed abstractions

Underlying attitudes connected to social norms relating to driving ability and masculinity are perhaps counteracting the message of this advertisement, in particular the tendency of drivers to believe that they are safer than the average driver and that safety messages do not apply to them (Walton and McKeown 2001), are perhaps counteracting the message of this advertisement. In these discourses males in particular, regarded themselves as good drivers, able to handle a bit of extra speed and did not consider speed a real problem. Females recognised elements of this discourse that were being addressed in the advertisement when the driver said 'sorry it's the car'. However, they were not able to challenge the speed issue more directly and to put forward reasons as to why speed was a problem in itself. In this sense they were also restricted by the underlying view that speed is not a problem in itself, which thus remains unchallenged by the advertisement.

Harré, Foster and O'Neill (2005) suggest that the self-enhancement bias of young males leads them to exempt themselves from the message of advertisements that show dangerous driving resulting in a crash. Young males take the message as not applying to them because they see themselves as good drivers. Many of the responses of young males to the advertisement examined in this paper illustrate that they consider crashing in the circumstances shown as an example of bad driving, boredom or tiredness and not a consequence of speed. A dangerous situation is one that their driving ability should be able to handle not one that they might bring on themselves through their choice of speed.

The attitudes evident in this discussion were connected to social norms of gender for both men and women. The men demonstrated confidence in their own abilities and a willingness to maintain that risky driving was not risky if you were a good enough driver. The women on the other hand expressed concern for others but this did not necessarily make them challenge the men's dominant views.

The argument of this chapter is that drivers are made not born and that driving cultures related to gender and age are strongly defended and maintained. Social and cultural forces that precede the individual play a significant part in shaping how a practice such as driving is engaged in. While individuals have to be held accountable for their actions, there are aspects of culture that need to be included in the accounting and challenged. Although combustion masculinity is a dominant standard related to cars there are other forms of masculinity expressed and practiced through motor vehicles. Hydraulic masculinity as a kinder, gentler form of masculinity will be explored in Chapter 6.

One of the most significant areas where combustion masculinity can be reframed or at least curtailed is in car advertising and promotion, which relies on the excitement and unpredictability of combustion power in advocating the need for powerful cars and the driving styles purportedly suited to them. Future research can examine the inclusion of values of aggression and excitement as they are inappropriately applied to cars in advertising as well as attitudes that are implicit in beliefs about driving practice.

Young men need to be educated to take into account the environment, particularly the social environment and the context of traffic and access to roads, non motorised forms of transport and pedestrians. Teaching young men to drive slowly with patience and care would be more useful than focusing on cornering as fast as possible. The relationship between masculinity and cars could be illustrated in different and more positive ways focusing on enhancing planning and observation skills rather than just handling skills. Most men are capable of taking more care in their driving if more emphasis is given to this aspect of driving rather than considering the car and driver as king with right of way over everything else.

Chapter 4

Driven by Desire

Desires are associated with sexual energy requiring an outlet and discourses such as advertising that appeal to desires promote a connection between cars, speed and male desire, thus creating a desire for speed. The commercial processes referred to in Chapter 2 involved in commodification, that is, turning objects into commodities, also involve the commodification of desire whereby desire which is concerned with achieving or maintaining states of pleasure or joy, is connected to objects of consumption.

The bodily pleasures of driving are implicitly bound up with desire in the practice of driving and relationships to cars. Traditionally, men have been regarded both as having a closer connection to the raw pleasures of speed and technology and being more rational while women are considered less skillful than men and more emotional. Explaining how men can be seen as having a greater need to pursue their natural desires and at the same time able to be more rational is part of the task of this chapter. In driving it seems that men could be more emotional and women more sensible because they are not driven by excessive desires for speed, danger and mastery. Research has suggested that women are more emotional as drivers in positive ways in that they experience worry and concern more than males (Rundmo and Iversen 2003) however different emotions are likely to be experienced by men. The ways men and women experience driving differ in the kinds of bodily and emotional experiences each has and these will be outlined through focus group comments in this chapter.

Road safety literature seeks to appeal to the rational driver while advertisers tempt the desiring driver (Morphett and Sofoulis 2005). The chapter will show how a continuity between the rational and desiring/emotional driver can be established that is more beneficial to understanding and addressing road safety issues than either ignoring the role of desire or opposing desire and rationality. Social theorist Barbelet maintains that there is an emotional basis for all action and that this is the source of motivation for rationality. Judith Butler in her recent book *Undoing Gender* (2004) considers desire as the desire for recognition, to be connected to others and to the social context.

The chapter will give an overview and assessment of the most prevalent theories of driver behaviour and focus particularly on research on the emerging role of the emotions in driver behaviour. The theory of planned behaviour has played a significant part in psychological theories of driver behaviour (Parker, Manstead and Stradling 1995). Other theoretical approaches have attempted to address the behaviour of young drivers through constructs such as 'recklessness' and 'sensation seeking' in developmental (Arnett 2002) and social approaches (Jessor 1992). Another important area in psychological theory is cognitive approaches and there has been some investigation of the application of cognitive accounts to emotion and driving behaviour also to be discussed (Rundmo and Iversen 2003). One of the themes that emerges most strongly from research on reckless driving in adolescence is the significance of the emotions and how they are managed in the context of driving. How the emergence of the role of the emotions impacts on the broader theories is discussed.

Driver behaviour and emotions

A number of researchers have looked at the role of the emotions in the risk behaviour of young drivers (Arnett 2002; Rundmo and Iversen 2003; De Joy 1992; Harré 2000). Intense emotion appears to be an experience of adolescence. The emotional engagement involved in driving needs to be considered more broadly however. Emotions, it is often assumed, are overseen by rationality in adulthood, but appeals to rationality are not necessarily successful in campaigns for safe driver behaviour. Among the social and cultural influences that need to be considered in appealing to people to change their behaviour, emotional engagement is one of the most important (Sheller 2005).

Figure 4.1   **Safety or speed message? Does this suggest that speeds now are absurd or is it asking people to imagine how fast it will be in the future given how much has changed since then?**

People's attachments and forms of expression in their everyday lives include and are expressed through cars in societies where the car is not only the dominant mode of transport but also seen as a very personal vehicle of expression that can help fulfil every desire. We do not merely shape the car, the car shapes us (Sheller and Urry 2000, 2003). Emotional engagement relates to the attachments and forms of interaction that people experience through and with cars, in living in a car dominated world. '... the social meanings of commodities ... are far more important to the ways in which our society works than the *utilitarian* meanings of these goods' (Maxwell 203). Maxwell emphasises the meanings of car use as embedded in the social relations of everyday life which are expressed in 'emotional attachments' to cars and the ways in which cars function in our lives.

For anthropologist Mary Douglas (1992) the risks individuals focus upon have less to do with individual psychology and more to do with the social forms in which individuals construct their understanding of the world and themselves. The social interaction involved in car use is significant in shaping the forms of emotional engagement that individuals enter into. There is an assumption in road safety that rationality determines what people do and appeals for safety are essentially appeals to people's rationality. Rationality in this sense is action determined by objective facts and not emotional responses. Aberg (1998) states:

> It is difficult to understand why traffic safety rules should be necessary to change behaviour. If they were behaving in a rational way, drivers who learn that their behaviour is associated with increased risks should avoid performing that behaviour and then traffic safety rules would not be necessary. However, there is plenty of evidence ... indicating that information about risky behaviours has very little effect on actual behaviour. (209)

It is believed that if people used their rationality more to determine their driving behaviour, the roads would be altogether safer, smoother and more efficient and that it is information that is required to appeal to rationality. An assumption implicit in this view is that everyone, when rationally motivated, is primarily interested in the safety and well-being of themselves and others.

The theory of planned behaviour, as expanded by Parker, Manstead, Stradling, Reason and Baxter (1992) in the context of road safety, often underlies approaches to road safety and understanding infringement tendencies, as a model of the structures that determine or influence driver behaviour. The model focuses primarily on rational understanding with some consideration of individual and social perceptions influencing decision making. The theory of planned behaviour will be discussed first, followed by a discussion of Rundmo and Iversen's (2004) distinction between cognitive and affect based risk perception. Accounts of emotionality in adolescence are then considered and emotional engagement with cars more generally. It is concluded that emotional engagement appears to be limited in the research to expressions of aggression and as a characteristic of adolescence, but should be expanded to understanding driver behaviour more broadly. How emotional engagement is shaped in relation to driving is significant in understanding driver behaviour.

Theories of driver behaviour

Planned behaviour

The theory of planned behaviour allows behavioural intention, made up of attitude to the behaviour and subjective norm, to be predicted. A third determinant of behavioural intention, perceived behavioural control, was later added in order to allow for behaviours which are not completely volitional in character (Parker, Manstead, Stradling, Reason and Baxter 1992). This refers to the 'degree to which an individual feels that performance or non-performance of the behaviour in question is under ... volitional control'. (94) It was found, in the case of four driving violations, drinking and driving, speeding, close following and dangerous overtaking, that as perceived behavioural control increased, behavioural intentions decreased (99).

The authors concluded that; 'Some respondents may have unconsciously underestimated their control over these behaviours to protect their self-esteem or social esteem.' (99) Nevertheless, the fact that drivers appear to explain their bad behaviour by perceiving that they have little control over it shows that by stressing the control that drivers do have, and the need to take responsibility for their behaviours, has implications for driver training and road safety messages, according to the authors. It should also be noted that the feeling of lack of control could also be explained by the influence of broader social and cultural factors on driver behaviour.

The study by Parker et al. showed consistently significant correlations between subjective norm and intention, rather than between attitude and intention. They explain that the importance of subjective norm is possibly due to driving behaviour being a 'social performance carried out in the public domain', and involving consequences for other people (100). Thus, a behavioural intention with important implications for others is likely to be formed with the perceived views of significant others playing an important part.

The behaviour of significant others however, may have an effect that is not concerned with caution towards others but rather expresses aggressiveness for example. Young men in our focus groups did not express concern about the effect of their driving on others, which was often expressed by young women:

> If you crash into another car, then you're hurting other people as well ... so I always drive carefully. Other people think I drive like a grandmother! (laughs) (Female 2, Goulburn 1)

The men on the other hand wanted to emphasise that their driving was good and it was other drivers they had to be concerned about:

> I don't usually take my car to go out. If I go out, I go with other people so I can drink. Plus, I'm scared of someone hitting my car! (Male 1, Bankstown)

This young man described himself as a 'safe driver' saying he did not have to worry about what he might do to his car. After a young woman described her nervousness when she first learnt to drive and how she got scratches on her car, another young man was keen to express the difference between the men and the women:

> That's the difference between us. What he was trying to say was, because he's got a new car, it's not that he doesn't drive better, it's that he's scared of other drivers ... because we don't want anything to happen to our car. (Male 2, Bankstown)

There is an explicit agreement amongst these young men that it is other drivers they have to worry about, not their own driving. Many young men said that they drove more carefully when they had others in the car, depending on who was with them and the situation:

> I'm a very safe driver when I've got other people in the car. It's when I'm by myself I want to test the limits! But when I've got other people in the car I'm a bit more conservative ... (Male 4, Goulburn 1)

> It all depends. Like you said, if I'm out with my grandma or something I'll drive at 80 kms because if she even gets whiplash it might kill her, but if I'm with someone like J ... , the faster you go the more he likes it! He doesn't mind if I put my foot down a little bit every now and again, where if I drive with my mum or something, as soon as I get anywhere near a hundred she's like, 'Oh, you're speeding! You're speeding!' To get me to slow back down again. (Male 3, Goulburn 1)

The behaviour becomes routinised whether it is legal or illegal, safe or unsafe behaviour and this is to some extent what makes people feel the behaviour is beyond their control. While they know they should be driving to the rules they have developed habits as a young person that have become an implicit part of how they drive and these are derived from the driving cultures around them. One young man articulated both the helplessness of habit and the understanding that there was no excuse for ignorance:

> I reckon it's just habit. Like, the way you drive, you're more inclined to break the road rules if you've already broken them before. Like, I mean, I don't think you can say it's ignorance. You can't blame ignorance, if you didn't know there was a speed camera there. If you're more inclined to speed a bit, then you always will. You can't change that. (Male 3, Fairfield)

Parker et al. (1992) found that younger drivers in their study perceived less pressure from others to avoid committing driving violations than older drivers, and were also more highly motivated to conform to the perceived desires of their peers. Younger people were inclined to evaluate the committing of violations less negatively, and younger men differed significantly in the degree to which they considered committing driving violations to be under their volitional control (100).

In their 1995 study, Parker, Manstead and Stradling investigated the role of personal norm in extending the theory of planned behaviour. Personal beliefs about what is right and wrong were not taken into account in the theory of planned behaviour, they argued. The construct personal norm is intended to reflect people's internalised moral norms. Social norms on the other hand, reflect people's perceptions about what others want them to do. Personal norm is made up of moral norm and anticipated regret. Anticipated regret they considered important because 'driving violations are (for most people) socially undesirable' (129).

They conclude the 1995 study with the statement: '… measures of personal norm provide a useful extension of the theory of planned behaviour model, particularly when the behaviour involved is such that moral obligation can reasonably be expected to have a role' (136). The assumption is that social norms on which moral norm is based would be in agreement with the road laws. In a later study Parker, Stradling and Manstead (1996) showed a further increase in the predictive performance of the planned behaviour model, with the addition of a measure of moral norm and a measure of anticipated regret, based on the assumption that committing driving violations is likely to invoke feelings of anticipatory negative affect.

In the testing and application of the theory of planned behaviour the focus has overwhelmingly been on the committing of violations and aggressive driving behaviours. Affect is regarded in an instrumental sense as contributing to moral assessment. The contribution of affect to everyday driving and to driver behaviour in general has been considered very little.

The importance of personal norm and the perceived implications of behaving in a socially undesirable way are significant in understanding driver behaviour. However, the assumption that most people consider driving violations undesirable is problematic. While drivers know the rules and the potential for enforcement, the rules do not necessarily determine what is socially desirable. Increases in speed and better roads to facilitate faster travel, have been considered socially desirable or perceived as socially desirable, in many Western societies. Many drivers regard travelling at 10km/h above speed limits as acceptable and thus not really involving committing an offense. These kinds of social influences are part of the emotional engagement with which drivers interact within the driving community.

Figure 4.2   **Mitsubishi Lancer desired by young men has been associated with fast paced rally driving**

The theory of planned behaviour does not take into account the range of emotional engagements involved in cars and driving, and being part of the driving community, even though it acknowledges the role of social norm and moral norm. The emotional make up of engagement with the traffic environment has not been explored a great deal in the road safety literature. The particular complex forms of social norm, informed by a range of attachments and forms of expression need some exploration in relation to cars and driving.

Cognition and affect

Some research has attempted to show the importance of considering emotion in driving. Rundmo and Iversen (2004) distinguish cognition and affect based risk perception as separable but related aspects of attitude. They argue that affect has a primary function in judgments since a shorter response time is required than for more 'complex' cognitive judgments. Cognitive judgments in the context of driving include assessment of the probability of an accident and therefore relate to risk perception. They found that probability judgments and concern about traffic risks, did not seem to be important for risk behaviour in traffic, whereas affect did seem to be important.

The focus on perceived risk in their study, separated into cognitive and affective components, showed that worry and emotional reactions significantly predicted behaviour: the more worried and 'emotional' respondents were, the less risky their behaviour in traffic. They suggest that measures aimed at influencing behaviour and enhancing traffic safety should be 'directed more at influencing worry and emotional reactions to traffic hazards' (18). This was considered to be even more important where respondents scored highly on sensation seeking and were reported to be normless or indifferent to traffic safety.

Figure 4.3   Mazda 3 has also managed to attract significant interest amongst young men who pride themselves on their ability to dart in and out of the traffic. This one had a modified exhaust making it easier to pick as a young man's car

Two points are important here. Firstly, the idea that emotional awareness is linked to less risky behaviour would seem to go against the grain. The type of emotion involved is perhaps the significant factor however. Feeling 'worried' could be seen as comprising an awareness of the dangers constant in driving, and a sense of vulnerability. Nervousness and 'worry' or fear are not generally considered advantageous in driving and are more likely to be related to lack of confidence. Young male drivers consider young female drivers as not as good because they are 'less confident' which is linked to the expression of more concern by the women. Young men are more likely to express confidence in their own driving and see this as indicative of better driving ability. Rundmo and Iversen note that males possibly relate less easily to emotions while females reported higher accident probability, more worry and reported emotional reactions more than young males. They report a need to develop measures to reach the young male group and develop emotion-based risk perception.

In considering the differences between males and females on perception of emotion an interesting concern arises which ties in with assessment of driving by males being based on their perceived level of skill and knowledge. The need for at least some males to see themselves as able to 'handle a car' and not feel afraid could be related to a need to overcome any feelings of vulnerability by privileging skill and this is related to norms of masculinity (DeJoy 1992; Clarke, Ward and Truman 2005).

Secondly, Rundmo and Iversen conclude that some drivers are 'normless', particularly those who scored high on sensation seeking. Rather than being normless, however, their norms are not those primarily important to traffic safety. They are possibly governed by another set of norms that are not necessarily alien to the road environment. Norms related to excitement are often expressed in advertising, taking the form of speed and aggressive driving. Norms derived in interaction with others are also significant, and when the social behaviour of young drivers is considered, norms that are not compatible with safety appear to be more influential. These norms are likely to be related to forms of emotional engagement and their associations that are not governed by safety concerns. The excitement of independence and freedom, and of being with friends is often associated with a carefree attitude and riskier behaviours, or at least a lack of awareness of safety issues as noted by Harré (2000).

Adolescents and emotionality

'Recklessness' and 'sensation seeking' have been studied as traits relating to particular behaviours in cars (Jonah 1997). Some individuals are seen as pursuing risk, that is, as actively seeking it out. Jessor expressed this as problem behaviour (1987) which he later couched in the context of the larger social context (1992). Risk behaviour he defines as, 'any behaviour that can compromise an adolescent successfully achieving psychosocial development' (378). Risk behaviours can serve social and personal functions, purposes, instruments and goals such as: peer acceptance and respect; autonomy; defying convention; coping with anxieties and fears; affirming maturity;

and adult status (377). He notes that adult behaviour sometimes exhibits similarities to adolescent risk-taking.

Feelings of invulnerability in adolescents are often mentioned and attributed to adolescent egocentrism (Elkind 1974; Cohn, McFarlane, Yanez and Imai 1995). There is an apparent lack of understanding of the implications or acceptance of the appropriate norms for safe driver behaviour amongst adolescents. Norms that are relevant to them or that they are likely to be grappling with include norms of individualism, defiance of authorities, masculine risk-taking, freedom as doing what one likes and feels comfortable with and consumption related to stimulation and excitement.

Young drivers' lower perception of risk reflects rather a 'failure to perceive dangerous situations than a desire to pursue risks' (Cohn, Macfarlane, Yanez, and Imai 1995). It is suggested in these accounts that adolescents are not emotionally mature enough or their actions are governed by their emotions where the behaviour of adults is governed by rationality, and knowledge of appropriate beliefs and attitudes. Cohn, Macfarlane, Yanez and Imai (1995) in their investigation of teenager's perceptions of safety and risk relative to adults concluded: 'Young drivers ... perceive less risk in tailgating, speeding, and night driving than do older drivers, which suggests that accident rates among youth may reflect a failure to perceive dangerous situations rather than a desire to pursue risks. (221) McKnight and McKnight (2003) report that of the 2000 non-fatal accidents involving young drivers aged 16–19 years, errors in attention, visual search, speed relative to conditions, hazard recognition and emergency manoeuvres were found to be influential factors in the crash rather than high speed and patently risky behavior which accounted for only a small minority.

Harré (2000) explores this point and, in addition, discusses the impact of intense emotion on the concentration of young drivers resulting in reduced risk perception. Temporary reduction in risk monitoring can be effected by distraction from passengers, loud music and intense emotions. She cites evidence from Schuman (1967) and Jung and Huguenin (1992) indicating that young drivers are likely to drive to 'blow off steam' to 'cool off' after an argument:

> The evidence cited ... suggests that drivers who experience intense emotions may have a relatively high crash rate and that adolescents may be more prone to some types of emotional driving than adults. (Harré 2000, 213)

Harré notes that it is not clear whether intense emotion acts as a distraction reducing risk monitoring or 'an incentive to engage in active risk seeking' (213). She concludes that in all likelihood it can operate in both ways, reducing risk perception in a young person who is overwhelmed by an emotion such as anger and consequently makes poor judgments, and in a second scenario, 'fuelling a desire to *take* risks such as speeding, cutting corners, and so on', and requires further investigation (213).

Arnett (2002) has looked at the sources of crash risk in young drivers in developmental terms. He distinguishes adolescence (10–18 years) from emerging adulthood (18–25) and focuses on the differences between 16–17 and 18 year olds in their driving related behaviour. The younger group have a higher crash rate and are more likely to be living at home and attending school. Their peers are the focus of their leisure activities and their judgments about how they look, their popularity and what they do. Cars become a means of socialisation freed of parents, and adolescents experience their happiest moods around their friends. The elation they experience in their independence, being with their friends, without adults, is expressed in extreme positive emotions that can have the effect of distracting and encouraging the driver to take risks. Arnett notes that adolescents report more extreme emotion than pre-adolescents or adults. Adolescents are more likely to drive to 'blow off steam' and express extreme negative emotion:

> It could be ... that one of the reasons for higher crash rates among 16–17 year olds is that they spend more time in cars with friends and use their cars more for purposes that promote their social interactions but are inimical to safe driving. (Arnett 2002, 20)

The importance of social interaction and the role of emotions are clearly significant, and likely to be more significant in influencing young driver behaviour than rational understanding of the risks associated with driving. This is not merely a factor of youth though it is related to the development of emotional control and expression. Social interaction and emotional responses, such as pleasure and boredom explored in Chapter 2, are likely to influence the responses of drivers in general to some extent.

In much of the research on young drivers, the social norms operating with young people appear to involve forms of engagement that are not primarily about safety but this is also the case in the broader community where other norms operate that are not conducive to regard for safety. There is certainly evidence that some are keen to absorb risk as a challenge to themselves in various arenas such as, skydiving, rock climbing and other extreme sports (Lyng 1990). In road safety however, the expression of risk-taking in cars, particularly for younger drivers needs to be seen in the context of the way cars are represented, as well as adolescent development. With advertisements which emphasise power and performance and show the car as allowing the driver to feel like a race driver, cars are interpreted as enabling the exploration of danger and excitement and thus as appropriate means the expression of risk behaviour. Adolescent behaviour needs to be examined in light of the kinds of emotional attachment and expression that are encouraged in and through cars as well as a consideration of the emphasis on social relations.

Driving as emotional engagement

The emotionality of adolescents needs to be seen in the broader perspective of social and cultural forms of emotional engagement in the driving community. The emphasis on some aspects of driving such as the thrill of speed, the focus on driving skill and making social exchange on the roads more impersonal, is not confined to young males. The social interaction involved in driving is expressed through forms of emotional engagement that young people enter into. The research focus on emotional engagement in driving has overwhelmingly been aggressive driving to the almost complete neglect of other emotions. Furthermore, anger is largely treated within a model of emotive versus rational behaviour rather than as a form of emotional engagement in driving.

DeJoy (1992) states that there is a need for more personal appeals in road safety, for interventions that 'personalise the risk', since most drivers perceive risks as not applying to them personally. Young males in particular are more optimistic, especially regarding driving skill. In the research he cites, males and females had similar perceptions concerning the frequency and likelihood of risky behaviours but males perceived behaviours as generally less serious and less likely to result in accidents. De Joy relates this to the expression of masculinity in our culture, as does Harré (2000). The emotional engagement involved in this perspective requires further exploration, and is likely to highlight an emphasis on thrill, aggression and excitement rather than an awareness of vulnerability. As Harré states; the 'social system of norms and media images [equates] fast driving and 'skillful' maneuvers with masculinity, adulthood and peer group approval'. These associations, she suggests, need to be dismantled in the long term. (2000, 218)

The well documented fact that drivers see themselves as less risky than other drivers (Williams 2003), possibly indicates a particular emotional engagement which avoids an awareness of feelings of vulnerability as Rundmo and Iversen (2004) suggest, particularly in males. An emphasis on rationality as opposed to emotion belies the emotional investment involved in attachments to cars and driving skill.

Shinar (1998) describes 'inconsiderateness towards or annoyance at other drivers (tailgating, flashing lights, and honking at other drivers), and deliberate dangerous driving to save time at the expense of others …' as '… a syndrome of frustration-driven instrumental behaviours' (139). These behaviours are not uncommon and comprise a structure of emotional engagement oriented towards aggression and competition. One young woman in a focus group remarked:

> Traffic just gets me frustrated. I hate driving in traffic. People are changing lanes and darting in front of you, and you've just got to be so aware of everything … like, it makes me more aware of things that are going on, but by the time you get home and you've been sitting in traffic for a while, you're so frustrated and angry … (Female 3, Goulburn 1)

Traffic is viewed as an everyday necessity that cannot be avoided perhaps contributing to the feeling of frustration and anger especially since the car is meant to be the vehicle of freedom and convenience. The awareness required in traffic seems to be experienced as frustration or alternatively the response of frustration leads to more awareness.

Research has also investigated the role of aggression, sensation-seeking and impulsiveness in driving (Dahlen, Martin, Ragan and Kuhlman 2005). Recent data has reportedly shown the prevalence of driving anger and aggressive driving (Neighbors, Vietor and Knee 2002). These studies have focused primarily on characteristics of individuals while considering the combinations and interconnections of the different scales employed, but the extent to which aggression has become a part of the social engagement of driving is outside their scope. There has been a spate of studies on aggressive driving in the last few years which have indicated that driving anger represents a 'significant and dangerous phenomenon', that is a 'frequent and more socially acceptable outlet of expressing negative emotion' (Nesbit, Conger and Conger 2007, 157).

A significant association was found between anger and aggressive driving in the review and analysis of the literature undertaken by Nesbit et al. but this did not correlate with a higher accident rate for angry drivers. This could be because other drivers compensate for the behaviour of aggressive drivers and move out of their way suggesting that a correlation could be found between aggressive driving and hostility or intimidation. The impact is more significant than crash rate in that it can have an effect on overall quality of life because other drivers and nonmotorised road users must allow for the most aggressive drivers. As stated previously the car lends itself to intimidation by virtue of the damage it can do and when combined with a willingness to use it to express anger it is potentially very dangerous.

Crash rate is perhaps not a good indication of the consequences of driving aggression. Nesbit et al. argue that driving could be seen as requiring a high level of endurance and perseverance and that arousal in the form of anger could enhance cognitive processing explaining the stronger relationship between anger and violations and the weaker relationship between anger and errors. 'Angry individuals may express aggression while driving, but they may also be more careful and attend to the danger presented by their actions.' (172) This is certainly an argument used by men – that increasing their arousal by speeding or other risk-taking activities makes them more engaged and aware, however there would seem to be some contradiction in claiming that drivers can be more aggressive and more careful at the same time. The equation of stimulation or arousal with aggression needs to be investigated and questioned as it is likely to be a result of broader social and cultural factors whereby arousal is considered only in terms of aggression. Additionally, such a conclusion does not take into account the impact on the social environment of driving aggression even when it is accompanied by more care in the form of skillful handling.

The higher incidence of male death and injury on the roads in part due to driving exposure has also been related to how males drive, as in exhibiting common risky driving behaviors (for example speeding and red-light running), males clearly out number females (Lonczak, Neighbors and Donovan 2007, 537). Lonczak et al. also note that males have been found to exhibit more aggressive driving behaviours such as confronting other drivers and road rage, defined as intentionally assaulting another driver with a motor vehicle or a weapon as a result of an incident on the road (536–537). Their study finds that while males are certainly more likely to receive traffic violations and to experience traffic related injuries/fatalities as compared to females overall driving anger was not differentiated by gender (541–542). Females were found to be more likely to experience anger when experiencing obstructions in traffic situations and in response to other drivers' illegal driving practices. The authors propose that the difference in female expressions of anger could be due to their unwillingness to break the law and the relatively greater acceptance of law breaking on the part of males (542).

Framing aggression

There are many reasons why aggression could be a significant part of the emotional engagement of driving, particularly for males, but a connection to the social framing of driving as a practice must be acknowledged and investigated. As discussed in Chapter 1 advertising contributes to social framing. The current emphasis on the need for 'mean' looking cars that provide safety in the 'concrete jungle' can be seen as contributing to the aggression associated with cars and driving. Muscle cars such as current models of tradesmen's vehicles like the Toyota Hilux, Holden Rodeo and Nissan Navara are designed to look like they can shove or intimidate others out of the way. The Toyota Hilux was marketed on television (2005) and on their website (http://hilux.toyota.com.au/TWR/NewHiluxFlash2/index.html accessed 14.01.08) with the tagline 'get in or get out of the way'. Similarly, Hummer shows on their website the substantial grill of the combat style vehicle as a set of almost sharklike snarling teeth, with the words 'Now Get Lost' (http://www.hummeraustralia.com. au/ accessed 18.01.08):

H3 – The Newest Action Hero

The HUMMER H3 has all the characteristics of a Hollywood action blockbuster. The ability to take the car almost anywhere makes every ride a thrill ride, and with a tough-guy attitude typified by our headline 'Now Get Lost', the H3 would not look out of place starring next to Bruce Willis or The Rock. (Available at http://www.hummer-time.com. au/?cat=6 accessed 18.01.08)

Hummer's television ad show the vehicle in urban contexts as much as 'jungle' and 'desert', firmly indicating that the vehicle is meant to appeal to city drivers and to an urban context where the tagline and angle of shot indicate that anything in its path that does not get out of the way will be run over. The appeal of the vehicle is meant to be its aggressiveness and in the urban context is intended to suggest having the edge through intimidation and power. The style and promotion of this vehicle contributes to the social framing of the driving experience as aggressive.

Another medium where significant social framing of cars and driving is evident is in motoring magazines. It is estimated that up to 38% of the Australian male population aged 16–34 years accessed motoring magazines (Sofoulis, Noble and Redshaw 2005 Appendix 1, 7) and in the Transforming Drivers media study it was found that 37% of media use questionnaire respondents reported reading motoring magazines (Sofoulis, Noble and Redshaw 2005 Appendix 1, 7). Motoring magazines often lament speed limits but rarely have road safety messages to put the enthusiasm for fast, powerful cars into perspective.

A high proportion of respondents in the media study, 90%, also reported having played driving-related games. Three quarters of the sample (60 people) mentioned having played games related to racing with cars, bikes and/or karts, and titles mentioned included *Mario Brothers, Need for Speed, Rally Championship, Gran Tourismo* and *Daytona* (9). Many of these games have aggressive speed and racing themes and can also be considered to play a part in the social framing of driving. In fact ten of the total of 80 respondents claimed that the games helped with navigating obstacles, steering manoeuvres, reaction time and hand/eye coordination. These responses tended to associate speed, skill and safety with the theme of control:

Yes, it can, by increasing reflexes and co-ordination.

Yes, car racing games assist you in terms of handling, driving and gear changing.

The game is helpful as it makes you concentrate on how fast you are going, keep an eye on other cars, especially when going around corners/sharp bends - have to balance acceleration with braking, steering, etc.

Yes. The more you want to win, the more you concentrate not to falter on the driving

Some respondents showed concern for how games could influence people in their ideas of driving:

Makes it worse I think, makes you think you can take a corner at 100 km/h and drop the handbrake, ending up fine. In real life you'd most likely crash.

I don't think so. Virtual reality driving (strengthened) would be more effective. Computer games perhaps increase reckless driving.

For others, the response was mixed. In the following quote the comment is made in the last sentence that it took some time to distinguish the game and reality:

No, since there is an element of 'shock' when driving real cars as opposed to virtual cars. The fact that your crashes are real, with real consequences, creates panic … it did for me, at least. It took a while to accept that I was no longer playing a game.

The form of driving experienced in the games is considered by some to be an appropriate model of driving on the roads developing necessary skills like fast

Figure 4.4   Men and boys play racing car games at Toyota's showroom in the Champs Elysée in Paris

cornering and navigation. There is nothing in these games to counter the racy driving style often reinforced in other media as the model of good driving.

Some focus group participants in the second Transforming Drivers study on gender expressed the view that the roads were a forum of aggressive competition and agitation:

> See, most of the time people do things just to piss you off, you know? They like being jerks. It could come from a lot of things. It could be jealousy, it could be whatever. Or they just want to annoy someone. You get them all the time, especially if you drive a nice car. Some idiot will always try to cut in front of you or do something silly. (Male 3, Fairfield)

The competition over 'nice' cars and driving them to demonstrate superiority takes priority over the car as a means of getting somewhere. It is as much a means of display as a means of being mobile, for some young men.

Aggression could also be a response to feelings of vulnerability in cars. People are generally wary of the driving of others and regard their own driving as 'safer' (Walton and Bathurst 1998). This phenomenon has not been investigated in terms of the vulnerability that people experience in cars and the ways in which they may respond to that vulnerability. The kinds of emotions associated with driving could be expressed in systematic responses that are socially acceptable as forms of expression amongst particular groups.

There do appear to be socially acceptable systematic ways in which people behave and others respond on the roads. One young woman explained how she felt when she experienced tailgating and understood what it was:

> I've been tailgated once before, and I was just horrified at how scary it actually was. I didn't realise what tailgating was until this person actually tailgated me. I don't tailgate people. I do get frustrated, but it doesn't reflect in my driving. (Female 3, Blacktown)

This young woman was experiencing for the first time the actual phenomenon she had heard about but never seen for herself from the driver's seat. It was a phenomenon she could recognise and respond to. She clearly got the message that this was meant as intimidation as she felt 'scared' and 'horrified' by it. It was more than having someone following closely behind but meant to suggest aggression. Many young women and some of the young men said that they would simply get out of the way of aggressive drivers or slow down so as not to give in to them.

A young man in another group explained how simple actions of others on the roads could make him angry and frustrated in the traffic because of the implications involved:

> I get road rage something fierce. The other day I was coming up Clinton Street and I was behind an L-plater and they were going quite slow, about 30 kms an hour, and I put my blinker on to turn left, and this hoon in a hunk of junk, it was, and I'm in the tangerine Ford and I'm thinking, 'The nerve!' (laughter) Anyway, he's coming up the side of me and I'm on the horn! He just looked at me like, 'Huh?' Anyway, I just went brrr!!, right up behind his backside to let him know! Then I dropped off. I'm not a total nutcase. (Male 2, Goulburn 2)

This young man was reacting to the slow speed of another vehicle as well as to another young man passing him despite the car he was driving. He was even indignant about the kind of car that he was being passed by – whereas he had an obviously newer and trendier 'tangerine Ford' the other guy only had a 'hunk of junk'. The implication was that the guy in the older car should be submissive toward the newer, superior vehicle/driver. There seemed to be a social etiquette involved that this young man felt should be complied with by others, especially other males. His response was to tailgate the other driver, and this was obviously meant to intimidate, 'to let him know' he was annoyed with him and that he had infringed a sacred rule. This seemed to him to be reasonable and acceptable behaviour and was clearly understood to be conveying a particular message.

New cars are promoted as faster and more powerful than preceding models allowing anyone to have the experience of the real enthusiast who loves the 'authentic' driving experience. Speed is seen as cool and 'macho' whereas slow driving is 'feminine' or 'old-fogey', and skillful, fast driving are the means to get ahead and to relieve stress and frustration (Silcock Smith, Knox and Beuret 2000). People can be annoyed or affronted by someone going slower than they themselves wish to, regarding the slower driver as deliberately obstructing them and responding by close following and flashing lights to intimidate them. Others may tailgate without noticing what they are doing but nevertheless invoke feelings of intimidation in the driver in front. Young males say that if another car overtakes them they have to overtake them back. It is an affront to them to be overtaken, suggesting to them that they are being seen as someone who does not know how to drive (Redshaw 2005).

These responses are seen as acceptable on some level since they often have the desired effects – the expectation that drivers will or should move out of the way of other drivers who are flashing their lights at them and close following is an understood action able to be interpreted by most drivers. Systematic emotional responses could be confronted and deliberated in positive ways so that drivers are able to focus on more empathic responses to other drivers rather than responding as if the actions of others were a deliberate affront to them. The focus on other drivers as deliberately trying to thwart our efforts is symptomatic of the individual rather than social emphasis of driving.

Studies have consistently found that tailgating, hornblowing and headlamp flashing are perceived as hostile and have given rise to high levels of anger (Parker, Lajunen and Summala 2002). Aggressive driving does not appear to be a rare occurrence. In a study by Parker, Lajunen and Stradling (1998) 89% of respondents said that they sometimes committed minor aggressive violations towards other drivers, usually to show annoyance. Behaviour of other road users is most often reported as prompting feelings of anger when driving (King and Parker 2008, 44) and it is the generalised tendency to judge other's behaviour in predictable ways that King and Parker are investigating in their study. They report that drivers who are high on trait aggression have been shown to be more likely to perceive hostility in the actions of others (44) and also that the degree of consensus between one's own attitudes and behaviours and those of others can be exaggerated (45).

The study found that respondents believed other drivers committed aberrant behaviours significantly more than they themselves did and the specific violations each committed were those they perceived other drivers as committing more than they themselves did (47). King and Parker suggest that this could be a means of justifying their own actions but it could also reflect the consensus among their own particular driving culture mistaken as indicative of overall driving culture. It could also be indicative of the general view of the roads as hostile resulting from regular reporting of road rage incidents and motor vehicle advertisers portraying the roads as a combat zone requiring aggression as well as people's own feelings of frustration in the increasingly dense traffic. It does suggest a tendency to see what one expects to see as King and Parker note, though where the expectation is derived from is likely to be broader than an individual tendency and to have some foundation in the social framing of driving and the expectations of the car as freedom machine, of open roads free from other vehicles and hindrances, frequently shown in advertising.

Literature on aggressive driving discusses negative emotions experienced in the context of driving (Dula and Geller 2003). There has a been a great deal of focus on negative emotions such as anger while positive emotions such as contentment and enjoyment could also play a significant role. How they are related to and experienced within the context of driving is important. Many drivers experience enjoyment in driving at safe speeds while other people feel annoyed when their free flow of movement, their 'enjoyment' of faster speeds, is 'obstructed' by other vehicles. Enjoyment could be experienced in a number of ways in relation to cars but is often associated with extreme speed. Joy can be experienced in being able to assist another person by taking them to an appointment or out for the day or being able to take the children somewhere different (Maxwell 2001). Even driving anger is interspersed through other moods encompassing emotions that may relate more to contentment or at least mild satisfaction. The negative emotions are related to beliefs or expectations about the way things should be, and the frustration of expectations, that is, the 'fiction effect' referred to by Morse, the split belief in which one momentarily sinks into another world knowing it is not real (1990, 193-194). The beliefs and frustrations it is argued here, are part of the emotional engagement with cars as are feelings of joy, contentment and excitement. Further discussion of the role of these emotions and their associations in driving could be fruitful in understanding influences on driver behaviour.

The extremes of aggression and nervousness appear to be aberrations of the normal response to driving experiences and are addressed as extremes. Nervousness is not the ideal driving emotion. There is more emphasis on being assertive and even aggressive as expressions of 'good driving'. Young men often refer to 'confidence' in their level of skill and ability to handle the car. Excitement and enjoyment are liberally fostered in car advertising in ways that could eventually promote aggression when enjoyment appears to be interfered with or obstructed. The associations of enjoyment and excitement are connected in the emotional investment in cars, to aggression.

Emotional engagement is more important to normal functioning in many activities than is often acknowledged. While emotion is regarded as exclusive of rationality and vice versa, the emotions are actually important aspects of how we understand and operate in many complex social situations. The major theories then, which are formulated to emphasise rationality and objectivity, need to be informed also by more 'personalised' responses to the road environment, encompassing emotional engagement. Given that every driver is responding in a form of emotional engagement with the process of driving, these aspects of the traffic environment should be taken into account. Understanding the emotional attachments involved in cars and driving in our society, could also help to inform traffic regulation in ways that are more significantly related to people's ways of operating on the road and the ways in which these are socially and culturally constructed.

Safety campaigns might appeal to the rationality of the driver and portray breaking the rules as wilful disobedience, while car advertisers appeal to desire, maintaining and drawing on the idea that males in particular are driven by the desire for high speed and being challenged by power and performance in cars. People can feel helpless in the face of desires and since desires can also be coming from somewhere else outside them, feel they have little control over them. Desires are understood as at once uncontrollable and having to be tamed by the values and structures of society when they are also generated in the context of the desire for social recognition and acceptance.

Safety campaigns can be used more effectively to frame driving as a social practice in positive ways emphasising the freedoms that are available through access to cars and the extent to which people do drive according to the rules. King and Parker recommend campaigns that are able to 'present both sides of the argument' (2008, 48) regarding the prevalence of driving violations in the general population. Presenting the extent of compliance could help to challenge arguments like 'everyone

Figure 4.5   Rally driving associations promote the excitement of testing one's skills and the aggression of competition. Toyota display in Paris

does 10kms/h over the speed limit' ('therefore I am just complying with general practice and keeping up with the traffic by doing the same thing') if compliance is in fact greater than noncompliance. Accepted practices do have to be related to a broader culture in which increased speed has clearly been regarded as a positive gain and expectations of continual increases in speed are prevalent, particularly amongst motor vehicle manufacturers.

Other aspects of the compliance issue also need to be addressed such as the view that compliance is merely passive and road rules are merely a guide for those who do not really know what they are doing, or in other words, do not have the skills of a race driver. The road rules can be seen as an invasion and negation of the freedoms gained through the car since driving is seen as an individual endeavour encompassing 'natural freedoms and desires' when it is highly dependent on social cooperation and a substantial social and public infrastructure. The opposition to regulation is often a battle between red blooded males who know how to drive and authorities who merely wish to obstruct and repress their desires. The issue of governance will be considered more fully in the next chapter. The concern here is that social framing related to cars and driving involves oppositions and implicit meanings that are embedded in driving as a social and cultural practice. The remaining sections of this chapter will discuss social and cultural approaches to mobility and what they can contribute to understanding emotional engagement in driving practice.

Social engagement

Cars are important tools of social recognition and expression, believed to be saying something about the owner/driver. New forms of social action produced through the car have been the subject of analysis in social and cultural research (Dant 2005). Aspects of the driver-car assemblage have become habitual and routine but it is nevertheless an engagement, and the emotional engagement of movement and being moved in relation to cars and places has shared, public and collective dimensions:

> Cultural styles, feelings and emotions underpin and inform the relationality of things and people in the material world. (Sheller 2005, 223)

The variety of feelings that are generated are 'neither located solely within the person nor produced solely by the car as a moving object, but occur as a circulation of affects between (different) persons, (different) cars and historically situated car cultures and geographies of mobility' (Sheller 2005, 227).

Barbelet's account of the continuity between reason and emotion draws attention to the importance of social recognition and the emotions that are the background of activity. His schema encompasses concrete experience as felt, as involving emotional response and emotions as socially derived:

> As persons in the same social situation will share a common awareness of their situation, other things being equal, *mutatis mutandis*, they will have a common emotional or evaluative reaction to it. In this sense emotions can be said to inhere in social relations. (Barbelet 2001, 71)

Though emotions form the background of activities they are often disregarded as having any real purpose or function since it is considered to be rational and logical understanding that determines action. In his analysis of resentment and class relations Barbelet reveals the complex connection between particular types of experience and the social structure. Emotion is shown firstly to arise from the structural relations of society, and secondly to be the basis of and giving rise to action. Analysis of structural relations according to Barbelet should include class inequality as well as trade and business cycle movements and cultural patterns. The action that transpires varies with 'the nature and distribution of the emotional pattern' and thirdly, 'these actions affect the social structure by either reinforcing current outcomes or leading to modifications in the relations between actions'. Emotion then has both a social ontology and a social efficacy and as a source of social action forms a link between changes in the social structure over time (Barbelet 2001, 81).

Desire is often considered to be the basis of actions in popular discourse such as the 'need for speed' where it is claimed that men in particular need to 'get it out of their system' by having available to them facilities and opportunities that allow them to express that need. Some would like roads with no speed limit, and drag and racing circuits present an opportunity to experience higher speeds. The idea that they can get the need out of their systems is based on a 'boil theory' of emotion where a pressure builds up and has to be expelled through a particular activity implying that it is a natural need that continually builds up and has to be 'let out'. As with 'combustion masculinity' discussed in the previous chapter, male sexuality is seen as a gushing force that requires relief and one popular and culturally accepted source of relief is through risk-taking and the thrill of speed in cars.

The purpose of this chapter is to question this idea of desire as a gushing force that must find release. In an alternative philosophical tradition stemming from Spinoza desire is seen as the fundamental nature of humans to persist in being but it is constituted in relation to social being. Judith Butler points out that this is the 'early modern precedent of Hegel's contention that desire is always the desire for recognition, and that recognition is the condition for a continuing and viable life' (Butler 2004, 236). Our existence as social, cultural and gendered beings is constituted through social norms, 'to persist in one's being is only possible on the condition that we are engaged in receiving and offering recognition' (Butler 2004, 31).

If there are no norms by which we are recognised then we are unable to persist in our being since we are fundamentally social and are thus dependent on the possibilities of recognition through social norms. These norms are limited and limiting in how they allow individuals to be recognised as human:

> ... our very sense of personhood is linked to the desire for recognition, and that desire places us outside ourselves, in a realm of social norms that we do not fully choose, but that provides the horizon and resource for any sense of choice that we have. (Butler 2004, 33)

Though they appear permanent and fundamental, these norms are changeable. It is not a matter then of whether there is emotion or norms governing behaviour but of which emotions and which norms.

The case of confidence

It has already been noted in this chapter that men often refer to confidence as important in their driving and a reflection of their level of skill. Barbelet's account of confidence and it's social role will give some structure to the involvement of emotion and the norms through which they are formed.

For Barbelet action is based on confidence in a possible future and confidence in oneself amounting to a willingness to act (2001, 82). The feeling of confidence is associated with bodily sensations of muscular control, deep and even breathing, and an overall sense of well-being. Barbelet relates confidence to other feelings and to a time perspective:

> Whereas depression is a morbid remorse over past events, anxiety is a fearful anticipation of future events; whereas security is a feeling of comfort in the present, confidence is a feeling of assurance about the future. (Barbelet, 2001, 89)

The emotional experience includes a cognitive element of appraisal or evaluation determined by relevance to the individual (84) but it also arises as a result of acceptance and recognition (86). Negative feelings of fear and apprehension are regarded as being less likely to lead to action while for Barbelet confidence promotes social action as well as 'going one's own way' (86).

In the case of driving, confidence has a particular meaning and implication for action. Cars and driving allow one to carve one's way in the world and demonstrating skillful manoeuvring of a car brings recognition and acceptance from others, particularly for males. Confidence here means feeling confident in one's own abilities as a driver and as a member of a culture of drivers. Confidence in oneself as a driver often relates to confidence in other things such as the ability to get where one wants to go, literally and metaphorically, to carve a place for oneself, to fit in as a male and to demonstrate one's prowess, competitiveness and aggression as they are significant to the social context.

Young men's confidence in their driving frequently amounts to feeling happy about their ability to corner and to handle speed but not necessarily as these conform to the rules, the social environment and the needs of others. The motivation to conform to social ideals is strong and these ideals provide norms or standards of what is acceptable, desired and recognised as appropriate. Male expressions of aggression and competitiveness, and the need to drive forward and dominate and vent their powerful desires are expressions of social and cultural norms. Males can be nurturing, caring, cautious and take their time to do things properly but these are not the qualities desired in males in dominant discourses.

Car advertisers do not show men putting a baby into their XR6s or SSSs and driving carefully so as to protect and nurture the baby. They usually show men by themselves in cars thinking of nothing but themselves and the sensation of driving with confidence. There are no cares or responsibilities in these advertisements and they show no concern for others.

The cultural emphasis on skillful handling can be seen in research relating to hierarchical models of driver behaviour. The first level concerns operational skills, the second tactical mastery of traffic situations and the third, the highest, relates to planning or strategic management of driving (Laapotti, Keskinen, Hataaka, and Katila 2001). Crashes relating to the operational level were found amongst the 28,500 novice drivers tested to be more common in the first 6–12 months of gaining a license and to be largely due to inexperience. Once vehicle manoeuvring skills improved, these kinds of crashes diminished. But there were also a higher number of failures at the higher hierarchical levels amongst the younger drivers in the sample (18–20 years) that could not just be explained by inexperience. The authors suggest this is due to the greater inclination to take risks amongst younger drivers evident also in the number of offences indicating a habit of disobeying the rules. Young male drivers had the highest accident and offence rates and were the more persistent offenders leading the authors to conclude:

> Repeatedly violating the traffic rules may indicate problems on the highest hierarchical levels of driving behaviour. The drivers' motives are not in concordance with safe traffic or the driver cannot control him/herself (e.g. willingness to speed). (Laapotti et al., 2001, 766)

Another study by Clarke, Ward and Truman (2005) looked at the development of vehicle handling skills in over 3000 accident reports involving drivers 17–25 years.

Figure 4.6   Customised older cars appeal to young men as their first car

They found that 'young driver accidents of all types are found to be frequently the result of 'risk-taking' factors as opposed to 'skill deficit' factors' (523). Further, they indicate that 'specific groups of young drivers can even be considered as above average in driving skills, but simultaneously have a higher accident involvement due to their voluntary decisions to take risks'.

While confidence in driving ability is mostly concerned with vehicle handling skills and does not include understanding of the complex social environment it is not an adequate basis for good driving. The hierarchical model shows the areas in which other skills are important in driving as part of a broader approach to and facilitation of living. Traffic and planning skills are clearly significant and require consideration and understanding of the social context beyond the driver both at the level of the individual driver and in understanding driver behaviour. The particular form that confidence takes in the context of driving culture then must be taken into account in considering the behaviour of young men in particular.

The connection of confidence to recognition and acceptance indicates the strong social importance of confidence and how it is structured in activities such as driving. Desire understood as concerned primarily with recognition and acceptance can be seen as channeled into the expression of confidence through cars, for men especially. Finding ways to deal with the need for recognition and acceptance is far more productive than combating the alleged male need to give expression and constant release to their apparently natural desires for speed and aggression. Rather than being a potentially uncontrollable force, desire can be understood as the fundamental desire to persist in being and to be part of a social context. The social context is demanding that men understand themselves as prone to the gushing force of their desires. Women are presumed to have an inbuilt desire to shop and have also been connected to the idea of desire as a gushing force in some car advertising such as an Alfa 2003 television ad featuring quads where one of them is driving the car and all experience an orgasmic feeling as the one driving flashes past exploding towers of water shooting up beside the road and speeds around tight corners.

Emotion is similarly held to be a gushing force that is at times barely controllable and at best a distraction from rational thinking however the thinking that accompanies emotion is often the fuel that keeps it alive. For example, where many drivers feel that other drivers are out to punish them, make their life difficult, show them up and so on, the response is going to be anger and aggressive determination not to be pushed around or conversely, fear and intimidation and a tendency to be pushed out of the way. If on the other hand it is recognised that most drivers are just like us, wanting to get where they are going, occasionally making mistakes, developing unhelpful habits and oblivious to any effect they might be having on us, we can be more generous towards them and perhaps prepared to consider our own contribution to the melee.

Confidence as it is socially framed in driving is connected to the idea of desire as a gushing force that can be let loose and reined in at will. Men's confidence in their skill relates to their measured expression and control of their desires in wrangling with the machine and demonstrating their prowess through it. Desire in this sense is aimed at nothing but a need to be vented. It is aimless and antisocial.

Theoretical approaches require a broader and deeper connection to the social and cultural involvement in individual driving behaviour and to recognise the implication of emotion, not necessarily as problematic in itself, but in particular forms that may or may not be helpful. The social and cultural context also comprises norms that are not those assumed to dominate or prescribed by the rules laid down by governing authorities. The norms governing behaviours such as driving need to be further investigated in the context of social, not just individual, engagement and involvement. Other ways in which social and cultural contexts promote dangerous behaviour and risk will be explored in considering the inconvenience of the car and the governance of driving in the next chapter.

Chapter 5

Dilemmas of the Car

We continue to put vast resources into car travel, with the enormous cost of fast and efficient roads upholding the value of our independence. We largely perceive that cars are 'safer' than public transport and the fact that they are more convenient is indisputable, but what kind of world are we creating with our passion for the car, what kind of convenience are we then favouring? The environmental costs are well known while the quality of life costs are just as significant. Ultimately the car cannot be disengaged from our desires and cannot be considered 'harmless' in any sense, but do the benefits outweigh the significant costs? In addition the type of person the car currently requires us to be, that is, aggressive, competitive and individually focused rather than socially responsible, is costly. While many approach driving through the sensation of comfort and familiarity, the increasing numbers and speed of cars on the road has been accompanied by an emphasis on combat skills on the road and the expression of the 'need for speed' while greater self-control is required in driving. Drivers may feel comfortable with a higher speed limit than the official posted limit and automatically drive at a higher speed but the system demands that they make the conscious effort to know what the speed limit is and to monitor their driving speed accordingly.

This chapter will first consider the inconveniences of the car that are often overlooked in our enthusism for it and the values it represents, before going on to consider the issue of governance and cars. Many of the contradictions in the norms operating around cars and automobility have been highlighted in previous chapters. It has also been noted that the ambiguities of the car are often over looked in the enthusiastic embrace of the dominant norms related to automobility. This chapter will aim to further draw out those ambiguities and multiple meanings that often work at cross purposes. The primary means by which the car/driver is governed it is argued here, is through the ideas of freedom, autonomy and desire that dominate in the context of western capitalist market societies. Those ideas of freedom are also the means by which governance through discourses of safety is made necessary however. Drivers are constructed as both rational and desiring and the conflict between the two is part of a social discourse of risk management. Advertising appeals to the desiring driver while safety campaigns appeal to the rational driver and these are considered to be opposed. Social discourses maintain an opposition between reason and desire, which is an implicit part of capitalist consumer culture (Barbelet 2001).

The inconvenient car

> ... there is a crucial importance in recognising that automobility is fundamentally political – that it entails patterns of power relations and visions of a collective 'good life' which are at the same time highly contestable and contested. (Bohm, Jones, Land and Paterson 2006, 4)

Motor vehicles have brought enormous capacities for travel and ease of movement but at the same time have created enormous problems. Air and noise pollution, particularly in cities is of significant proportions, but of greater impact is the violence brought to the streets through the domination by motor vehicles and their capacity for speed. Higher speeds increase crash rates as well as reducing quality of life and the habitation of streets. From the comfort of the car many drivers are unaware of the impact of their passing vehicle on those on the sidelines.

Creating the need for enormous parking areas and streets densely lined with vehicles, motor vehicles make streets and public spaces unfriendly and ugly places. Shopping centres surrounded by enormous car parks often make little allowance for entry other than via a motor vehicle. We can place the constant hum of traffic in the background of our consciousness and accept the necessity of car travel for most purposes thereby overlooking and remaining untroubled by the ugliness and invasiveness of automobiles. The cost to our quality of life remains nevertheless.

The fast pace at which many drivers travel small suburban streets where children are playing and people are walking creates a sense of violation and aggression.

Figure 5.1   Plenty of space for cars in suburban streets in the United States

Everyone quickly and anxiously moves out of the path of motor vehicles for fear of getting hurt and this has to be taught to children as early as possible. They have to be taught that the road belongs to the motor vehicle and that there is a boundary between road and footpath that they must be able to 'read'. They cannot leave the garden gate or even the front door without being aware of and learning to live with the enormous invasive power and danger of motor vehicles:

> The 'structure of auto space' (Freund 1993) forces people to orchestrate in complex and heterogeneous ways their mobilities and socialities across very significant distances. (Sheller and Urry 2000, 744)

Because other options have not been developed to the extent that the car has we become accepting of the limitations it places on our lives. In Australia, train stations often have extensive car parks attached to them so that people drive the few kilometres in city suburbs to the train and leave their cars for their return in the evening. In other countries such as Holland, spaces around train stations will be taken up with thousands of bicycles which people are accustomed to ride even in the cold winters, with children on the back and in their business suits. This is rare in Australia and there are few places to accommodate bicycles near transport centres.

As Sheller and Urry (2000) argue cars are both flexible and coercive, creating the need to manage tiny fragments of time while extending where people can go and what they can do and necessitating the flexibility and rights that automobility requires (743–744). While most young people enthusiastically embrace the car as

Figure 5.2   Bicycles at Eindhoven station in Holland

soon as they are able, they soon become aware of its limitations. Driven by the fantasy of unhindered movement, the realities of traffic and the need to feed and maintain the car can strike them as unexpected, frustrating and unreasonable. Many of the young people in the focus groups considered the car the most convenient and comfortable option for mobility, however, many also expressed the limitations of the car. Convenience has its own particular meanings for these young people as demonstrated in Redshaw and Noble (2006). Young people reported going straight to the drivethrough at McDonalds as soon as they attained their licence. This was a significant moment for them as being able to drive allowed them access to places that can only be accessed by car such as the drivethrough at McDonalds. Here the contradictions and self perpetuating nature inherent in car networks become apparent as the drivethrough was created for the car.

Convenience and ease of individual mobility are central marketing themes, especially for new technologies (Markussen 1995, Shove 2003). This convenience though masks the inconveniences such as driving around looking for parking which is favoured over waiting for public transport. Even when the private car is left at home the cost of catching a taxi is absorbed rather than the supposed inconvenience of public transport.

Opportunities are presented by access to a car that would not have been there without it and this in turn makes the alternatives seem uneventful and stagnant:

> We're never in the same area. Like, every week we travel. We go bowling in Mascot or to a club in the city or for coffee in Burwood. We're always somewhere different. If we didn't have a car, we'd be watching movies at home! (laughs) (Female 3, Bankstown)

Because these young people could expect to go a variety of places around Sydney to find activities, they considered not being able to do so as negative, as an absence of activity. Anticipation was created in those who were younger and looking forward to the chance to go further afield:

> All my friends are seventeen, and I can't do anything. I'm eighteen, right? And I can't go out or do anything because they're all younger than me. We have to wait till next year. (Female 2)

> So you have to go to the movies every weekend! (Female 2)

> Exactly! Precisely! I can't go out anywhere! (Female 3, Bankstown)

Being excluded from the activities slightly older young people are involved in meant she could not go 'anywhere' or do 'anything'. The sorts of social activities these young women from Western Sydney liked included going to clubs but also comedy shows or the drive-in and one liked to just 'drive off somewhere'. Although there were activities available in their immediate area these were considered boring and not as good as if you could go off to other localities.

Young women and men talked about driving to different people's houses 'to see what's happening'. Access to cars allowed them to go to more of their friends and continue on further, go into town and so on, but sometimes they would go to someone's place and 'just hang out' there. Many would share the driving, taking as many as possible in one car so they could all be together, to save money on petrol and because they prefer to go in a car with others rather than drive themselves.

Going out on the weekends and wanting to drink presented a particular problem about what to do with the car for young men from different groups. A young man from regional Goulburn explained:

> … if you live out of town, you come in, you drink and you have to leave the car wherever you put it, then you've got to try and find somewhere to stay, then you've got to try to get back to your car when you're sober again!

Problems with the cost and availability of parking were other concerns and the possibility of the car getting damaged were a source of unease to young men who had spent a lot of time, effort and money on their cars.

> Yeah, we go out. See, I don't usually take my car to go out. If I go out, I go with other people so I can drink. Plus, I'm scared of someone hitting my car! (Male 1, Bankstown)

Others highlighted the habits having access to a car got them into such as driving everywhere when they could walk. While a number of the young people mentioned walking to friends close by or to the shops, they were in the minority. One young woman from Goulburn explained that when her car was being repaired for several days her mother picked her up and drove here to work even though she was just around the corner; 'I'd die without a car! I'm too lazy to walk now!' Others in the group agreed and one young man exclaimed; 'I drive to the shop to get a can of Coke!' The can of Coke being small and insignificant put pay to the argument that the car is needed to carry all the shopping, a frequent justification for needing to travel by car. The young man was showing up his own habitualness by exclaiming that he would not even walk when all he had to carry was a can of Coke. Another theme implicit in the tendency to drive to get a can of Coke is that there is always the possibility that something else might happen or that once there it might be necessary to make other purchases. The car creates the need for greater consumption and a can of Coke also represents the possibilities of other things related to consumption as well as the unexpected.

The real costs of the car are often overlooked. An exchange in the Bankstown group showed how comparisons between the cost of public transport and cars often favour having a car because the real costs were invisibilised:

> I know for a fact your car costs less than public transport. I fill up twenty-five dollars every week, and if I want to go by public transport, that's going to cost more than that, because from my house to the station, I've got to catch a bus then a train. So it's going to cost heaps more …

A young woman in the group noted 'then there's insurance and rego' and the young man responded that that was 'only one thing' and 'it's only once a year'. The young woman persisted nevertheless:

Yeah, but it's still expensive. I just went through it all, and I had to get a few things fixed, and in that two weeks I've spent a thousand dollars on my car that cost me three thousand five hundred dollars!

Here recent expenses brought home to her the real cost of the car which can often be overlooked in order to maintain comfort with the habitualness of car travel. On one level we know there are good reasons to choose public transport but the car generally looks more enticing and fits the values we have promoted of privacy, freedom and individuality. Public transport is constructed as against all of these values, taking away freedom and independence by making us reliant on timetables and networks, passive to the mode of operation of large mass transport operations and removes our privacy.

Another woman claimed, 'If I can't drive, I can't work'. She went on to explain:

When I got this job, I didn't have my P's and it took me an hour and a half on public transport, and you know how annoying that is. I'd rather be in my car for an hour in traffic, listen to music, relax, not worry about whether I'm going to fall asleep and miss the stop.

The preference for traffic over public transport came down to the private space of the car being considered more relaxing in contrast with the possibility of falling asleep on public transport. The passivity of sitting in traffic was preferred over the 'annoying'

Figure 5.3   Car park at a station in Australia where commuters travel up to 75kms by train to the city

passivity of sitting on a train or bus. The exclamation 'you know how annoying that is' exposed the general agreement with such a view of public transport.

A young woman in another group noted that train travel was cheaper but it was 'unreliable':

> I could catch the train but it takes too long, and then if they're running late, or delayed ... (Goulburn)

She did say however, 'if we go out in Sydney we catch the train or the ferry ...'.

The inconveniences of the car are often overlooked because we are habitually used to automatically turning to the car instead of considering other options. Even a small walk can seem a long way when people are unused to walking any distance and have no sense of what it will take to walk or cycle.

Other factors worthy of consideration here include the neglect of public transport by governments in favour of the car and the proliferation of private spaces such as shopping malls that cater to cars and consumption and not the gathering of people for socialisation. Public spaces are not as developed as they could be because of the focus on the car and the promotion of consumption. Consequently, groups of young people who gather in shopping centres to 'hang out' are often moved on because they are regarded as a threat and taking up space that is devoted to consumption. The tendency towards private rather than public spaces has removed a sense of safety in public places and the opportunity for young people to be somewhere doing nothing but 'hanging out'.

The car is promoted as safe, cosy, flexible and powerful thus appealing to the values of freedom, independence and privacy in the particular forms that encourage private mobility. These values are not the problem in themselves rather it is the forms in which they are represented in consumer society creating constant desires for further fulfilment that habituates us to goods such as the car. The car has consistently been promoted to advance the desire for speed as a freedom and the right to go as fast as one is comfortable with as a fundamental expression of freedom as well as the particular privacy and individualism that it purportedly engenders.

The dangers of driving are often elided in the general preference for the private car, and the dangers of other forms of transport exaggerated. Many young people maintained that public transport was not only inconvenient as opposed to the convenience of the car, but also that public transport was unsafe, especially at night:

> I just like driving into the city, if I have to look for parking, I'll find it. It just feels safer. I don't like to catch public transport at all. (Female 1, Bankstown)

One young man put the ambiguities of travel by both car and public transport very well, recognising safety as an issue in each context:

> Public transport, when you're going by yourself at nighttime, whenever, you're in a train and you're thinking two things: I'm safe but I'm not safe. Someone could come and rob me and bash me. But when you're in a car you feel safe because you're driving it yourself, but you don't know what's going to happen, say, in two or three minutes. Someone might

hit you, or you might hit someone. So, it's both the same. You can feel safe at one time but you can't feel safe at the other time. That's how I put it. Like, when I drive I feel I'm safe when I drive because, to myself, I think I can drive. I know how to drive, I know my car, I know what the car will do, I know how to control my car. But I don't know how to control someone else who could hit me. Someone could come out the blue and bang, hit me and there I go. So, in public transport it's exactly the same thing. Anything could happen. (Bankstown)

The unpredictability of the roads is compared with the possibility of being assaulted on a train. Studies comparing the relative safety on public transport with that of cars by looking at accident versus assault rates need to be carried out. It is clear that there are far less casualties through crashes on public transport but even taking into account assaults it is likely that public transport remains the safer option. Nevertheless the car will be regarded as safer by most people because it is private and no one enters it unless the driver allows them to. One young woman in a focus group did comment that she drove around at night with her doors locked but the general expectation is that others only enter the private space of a car by permission, whereas we have little control over who enters a bus or train carriage or even who sits next to us.

Control is a significant issue in the preference for cars as alluded to by the young man above, since many feel that their own driving is good enough to control what happens to them on the roads. The car gives the feeling of being in control of one's own time and destiny and other modes of travel have traditionally been considered passive by comparison. The most significant factor in individual control that the car appears to facilitate is control of speed. Young people considered that if they were late they could just drive faster and get where they were going on time and they are not the only ones who maintain that speeding saves time. In urban areas in particular, very little time is saved by speeding as traffic lights and other hold ups will slow progress. Yet the effort of trying to speed gives the impression that one is doing everything one can to get to where they are going on time and making up for 'lost' time or reducing the wasted time that travel presents.

Because we are able to travel much further and faster we create the demand to do so. A days work may involve meetings across a wide geographical area as government departments and businesses draw together different sectors of their organisations that spread across and between towns and cities. Many organisations have policies that are meant to be reducing the amount of time employees spend travelling in order to reduce their exposure to risk. At the same time however there is a tendency to expect staff to cover large distances in one day attending different campuses to teach in the case of universities for example, or managing widely disparate parts of an organisation requiring extensive travel to meetings.

Travel in many of these cases is regarded as a waste of time as meetings are scheduled with not enough time allowed for travel between locations. Time is money and travel takes time so the effort to reduce that aspect of every kind of organisation is ongoing. Current discussion in Sydney about reducing speeds in some highly built up areas from 50kms/h to 30kms/h has produced responses relating to how much it would slow people down even though the average speed in the traffic in many areas is lower than that at busy times. It has been argued that people have to get their kids to school or childcare and then get to work and they do not need that to be made any more difficult by being slowed down. Their feeling is that they are able to speed things up within the margins available to them. It can be counter argued that the quality of life that would be gained by slowing the traffic down would be worth much more. The emphasis on speed and control makes that a difficult point to accept.

Time savings in the immediate future are unlikely to occur through physical travel so much as through technologies such as video conferencing and using the

Figure 5.4   Cars parked for miles around the opening of the first shopping
centre in Sydney at Roselands in 1965. Photo by Ken Redshaw

internet for skyping. Nevertheless the feeling of being in control remains when it is possible to speed in a car even a bit here and there. The sense of control that cars provide is again more limited than people experience or accept. There are real senses in which control is possible with a car such as leaving when you want and going where you want by the route you choose but there are many constant constraints resulting from the system that makes those choices possible. The next section will focus on the issue of control and how driving is governed.

Governing cars

Rules regulating traffic and driver behaviour have necessarily emerged with the evolution of the car. O'Connell argues in his cultural history *The Car in British Society* (1998) that class played a large role in both the need for regulation and the tempering of road safety demands by motoring lobby groups. In the late nineteenth and early twentieth centuries it was only the wealthy who had access to cars and there was a tendency on the part of motorists to flee the scene of fatal or serious accidents, many of which involved pedestrians – a minimum of 45% between 1926 and 1939, 63% when cyclists are included (O'Connell 1998, 116).

It is hard to imagine today that publications such as *The Economist* and the *Daily Telegraph* were critical of cars and yet the *Daily Telegraph* in 1903 ran a campaign against the 'social juggernaut' and *The Economist* noted in 1913 that '"the vehicles of the rich kill and maim far more people than the vehicles of the poor"' (O'Connell 1998, 119). O'Connell notes that it was the car that was the focus of criticism and not drivers due to the potential for it to become a point of conflict that the wealthiest owned cars while the poorer majority did not.

With the real possibility of class conflict looming over the emergence of motor vehicles various scapegoats were created beginning with pedestrians and then progressing to 'speed fiends' and others characterised as a danger and not part of the 'brotherhood of the road' (120). Most of those who owned and drove cars could then regard someone else as the problem and not themselves thus directing criticism away from cars. The criticism could then be directed away from cars. Women became a major target as drivers and have continued to be castigated and scapegoated right up until the latter part of the twentieth century.

The road safety solutions that began to be formed were moderated to suit the motoring lobby groups, which included in Britain the AA formed in 1905 'with the primary aim of frustrating police speed traps' (116):

> … a new science of road safety, concentrating on the education of all road users rather than further restrictions on motorists, gained credence amongst influential sections of society. (113)

The 'significant minority' who owned cars in Britain by the 1920s were influential because of their relative wealth and professional and commercial positions (123). The National Safety First Association (NSFA) in Britain was considered an independent organisation and came to be regarded by the government as the official body for the dissemination of road safety propaganda however, 'its agenda was barely distinguishable from that of the motoring lobby, who provided the NSFA with much of its finance and many of its officers' (117–118).

Speed has been an issue that has been fought over since the advent of motor vehicles and has mostly favoured the demands of the motor vehicle lobby. Thus for example speed limits on major roads were disputed by motoring groups from the earliest appearance of restrictions. Regulations relating to speed were necessary to deal with the consequences of the car. In the United Kingdom the *Locomotives on Highways Act* of 1896 removed the regulation that required a person to walk in front of a motor vehicle carrying a red flag and the speed limit was raised from 4 to 12 mph. The speed limit was raised again in 1903 with the Motor Car Act (O'Connell 1999, 13). Rising numbers of casualties combined with demands for higher speeds. In 1930 the 20 mph limit had been abolished as it was 'being almost universally ignored' (100). In 1934 a new limit of 30mph was imposed which had been fought by the NSFA despite intense public concern at the increase in casualties following the 1930 abolition of a speed limit (125). New arterial roads enjoyed by motorists attracted a 30mph limit that they resented:

> They felt that since their motoring taxes had paid for the roads, subsequent ribbon development should not entail an end to their freedom to travel at speed along the new highways. (142)

Some major roads were derestricted following petitioning from motoring groups (140). Resident groups mounted spontaneous revolts against derestriction organising demonstrations that stopped traffic. Mothers protested about the dangers to their children and cars travelling at high speeds failing to stop at pedestrian crossings (142). The motoring lobby countered by calling for boycotts of shops and businesses in areas where derestriction was being resisted. Meanwhile casualties continued to rise. The strength of the motoring lobby and the growing ownership of cars amongst professionals such as the legal community meant that there was very little legislation invoked apart from that already mentioned. Pedestrians were increasingly held responsible for road accidents, magistrates were reluctant to pursue prosecutions related to motoring offenses and the police were in an 'impossible position' having to enforce a law that was not supported by the 'general body of public opinion'. O'Connell notes that in this instance public opinion included motorists, their representatives, manufacturers, magistrates and police but not victims of road crashes or their families (135). Emphasis was placed on education rather than legislation and propaganda to educate all road users, particularly pedestrians, was relied upon instead of further restrictions of motorists (136).

The growing manufacturing industry in Britain added to the move towards facilitating motor vehicle use whereas in countries where there was no manufacturing, such as the Scandinavian nations, the earliest and toughest legislation emerged including the imposition of accident liability on car owners. Legislation for insurance and the compensation of road victims was introduced in these countries as early as 1918 (O'Connell 1999, 118).

O'Connell's argument is that road safety discourse replaced any real efforts to control motoring and spread responsibility amongst the non-motoring as well as

Figure 5.5   **Pedestrian refuge for crossing a busy highway between traffic lights in Australia. With traffic speeding past often above the 60km/h limit the pedestrian can feel very vulnerable while waiting for a suitable break in the constant flow of traffic**

providing scapegoats in order to preserve the idea that most drivers were gentleman. Graeme Davison (2004) made a similar case for Melbourne in Australia arguing that the motoring lobby was allied with the freedom to go one's own way and anything that got in the way or slowed the free flow of motor vehicles was seen as 'both hindering personal liberty and denying a law of nature' (116). It was considered that pedestrians should be educated and regulated so they did not obstruct the traffic and motor vehicles should determine the pace of traffic. Road conditions and slower moving forms of transport such as horsedrawn vehicles should be controlled to 'accommodate the speed of the car' (116).

Davison equally rejects the idea that motoring organisations such as the AAA were not political. In Australia as in the United Kingdom the AAA sought protection and advancement of the interests of motorists and lobbied parliamentarians accordingly, fighting taxation and petrol rationing and arguing for road funds. Since the 1920s the motoring lobby has allied itself with the conservative Liberal party of free enterprise against the Labor party which sought greater controls and regulation (118):

> In their pioneering years, motoring organisations in the 1920s and 1930s had constructed a political ideology around principles of laissez-faire liberalism, asserting the motorist's

unfettered right to the freedom of the road against onerous taxation, officious policemen, careless pedestrians and inefficient public transport. (119)

Governments in the United Kingdom and Australia tended to be susceptible to the pressures of the motoring lobby which worked tirelessly on all three levels of government in Australia. At the state level restrictive road laws were fought and public transport increasingly undermined, and at the local level successful lobbying achieved ever greater shares of public and private space (119). Public transport systems were gradually degraded through lack of investment while motor vehicle owners considered that they were paying their way. Massive public investment in roads and road networks has taken funding away from other means of transport and the total cost has never been truly borne by the motorist. There are also the additional costs of police for law enforcement, traffic regulation and surveillance and emergency services required by the proliferation of motor vehicles.

It has been asserted that government in the United States was not as involved in shaping the economic, social and physical landscape being created by the car as European governments where regulations governed the use of automobiles from the early days (Volti 2004, 19). The lack of regulations however, served to increase mobility across the vast distances of the United States enabling settlement of 'mobile populations to accompany economic investment in transportation systems' (Packer 2003, 143). Packer explores the links between personal mobility, governance and safety arguing that governing mobility is an implicit part of governance in neoliberal capitalist states. He notes that being a good American involves owning and driving a car and driving is sacred to Americans (137). Packer states strongly the thesis that mobility through cars has influenced the ideals and values fundamental to modern capitalist states: 'The American popular truths of freedom, wide open spaces, and individuality have been integrated with and altered by ... mobility' (138).

The promotion of speed and its association with freedom is in this view the product of particular forms of government rather than an innate need in individuals (139). Greater mobility has been necessary to the expansive demands of capital and has at the same time appeared as an advantage to individuals and an expression of freedom. Packer argues following Rose (1999) that freedom has become a resource for government rather than a hindrance to it (140). Freedom has become a means of governing and this can be made evident in considering how mobility is governed:

> A highly mobilised, yet less than nomadic, population is needed in order to adapt and migrate from old urban centres to new jobs in suburban sites of production (Packer 2003, 143)

A highly mobilised population is needed for the increasing numbers of jobs located in areas requiring commuting by car. Industrial sites are often placed near major roads such as motorways but not near public transport networks. This has been the case around Sydney and Melbourne where increasing development of road networks has linked in with the development of industrial sites further from city centres. Public transport networks have begun to service some of these areas but they are primarily located near motorways that allow ease of movement of goods into and away from the city. A mobile population is required by the capitalist system and its changes and movements to be able as drivers to transport their own time, money and labour.

Packer focuses primarily on the construction of safety as a discourse governing mobility however it is equally the case that risk and speed are part of that discourse. Safety and risk are produced discursively, that is, drivers and riders come to desire safety and/or risk and are versed in the practices that accompany those discourses. Speed and the production of anticipated increases in speed in particular have had a significant role in the governance of mobility since time equals money and speed is considered to save time, so that increases in speed are consistently sought and justified. The desire for speed is generally considered a natural desire that has to be restrained and repressed. Packer highlights the repressive hypothesis of police forces evident in police manuals which state that enforcement is 'generally repressive' and suggest that drivers and riders freed of surveillance and regulation would lose any sense of responsibility and let loose their deepest desires, leading ultimately to chaos (148).

Policing governance

Police forces have had a mixed relationship with the motoring public, at times having to enforce laws that nobody abided by and that in the end they disagreed with, but also having to deal with the carnage created by motor vehicles. In Britain the motoring lobby's preference for propaganda and education over legislation appealed to the police and the courts who were having to deal with prosecutions of ever increasing numbers of motorists (O'Connell 1999, 126). Attempts to bribe and intimidate police, to accuse them of infringements of liberties and claim that they should be 'protecting property rather than bothering law abiding motorists' were common in the period up to the 1930s (132). In the late 1930s policing on the roads began to change into an advisory rather than censoring role and then to education rather than prosecution while the courts became more and more sympathetic to motorists (133).

Policing has become more about enforcement in the latter part of the twentieth century and laws relating to motor vehicles have been tightened and strengthened however there remains a tendency on the part of police to favour a particular view of driving that is aligned with the combustion masculinity discussed in other chapters. Police recruitment advertisements shown on television in New South Wales in 2005/2006 have shown a police car sliding in dirt to convey the thrill of driving involved in policing. Police drivers are given special training on off road courses where they learn to handle skids and slides.

The police hold the view that they must have the driving skills to deal with police pursuits and emergency driving in traffic. Meanwhile safety discourses attempt to convince drivers of cars and trucks that no matter how good their driving skills or how quick their reaction times, they are not able to defy the laws of physics which control how long a vehicle travelling at speed will take to come to a stop. The same laws of physics apply to police no matter how much driver training they receive and while everyone will attempt to move out of the way of a speeding police car, especially pedestrians, provided they hear the siren or see the fast approaching display of flashing lights, there are still going to be casualties from vehicles travelling at high speed particularly in dense traffic areas.

In a cultural climate where constant stimulation is promoted and demanded the police attempt to show policing as an excitement filled career in order to attract new recruits. The freedom and permission to drive cars outside the usual restrictions is seen as one of the major attractions. There are contradictions apparent in claiming that it is alright for the police to drive cars as though they were not subject to the laws of the road or of physics because they are highly trained drivers but it is not alright for other drivers to undertake training for themselves and to see themselves as similarly equipped to handle emergency situations. Young men in particular like to display their skills and compete with the police by attempting to beat them in a pursuit for example. Research has shown that those drivers who have the most highly developed vehicle handling skills are often poorest in their judgements of appropriate behaviour on the roads:

> ... specific groups of young drivers can even be considered as above average in driving skills, but simultaneously have a higher accident involvement due to their voluntary decisions to take risks. (Clarke, Ward and Truman 2005)

This does not only apply to young men however. Police are seen as the arbiters of speed on the road and this is disputed by those who regard themselves as equal to or better than the police in the level of skill they have developed and in the kind of car they drive.

A video appeared on youtube in 2007 showing an Australian police car being driven unofficially at a local show in a display of lights and sirens doing burnouts in the dust. The accompanying discussion showed the ambiguous position of the police with some claiming that it was a great display and showed that police were human too and others that it demonstrated double standards since the police would confiscate the car of anyone else who did this. In one comment it is claimed; 'You cannot respect laws that have this draconian effect in isolation, let alone when you see police officers carrying on in this manner.' Another maintains; 'Well at least he's having a bit of fun with the community, building a rapport you could say. Cops these days are out of touch with the community.'

One person thought that the activity was tantamount to 'reckless endangerment' which they defined: 'A person commits the crime of reckless endangerment if the person recklessly engages in conduct which creates a substantial risk of serious physical injury to another person.' Some maintained that because this sort of activity was part of police training they could not see what all the fuss was about:

> Looks like they were just having a refresher on their defensive driving course. Pretty silly all the fuss when this is part of police training and because its on dirt nothing would have been harmed.

> I think idiotic would be if a cop didn't know how to control a car. If they're in a pursuit it's more than likely that the car they're driving will lose traction and oversteer, and if they

don't know what to do about that, I'd say that's much more idiotic then not knowing what to do in that situation.

Concerns were also raised about the activity being carried out where there were clearly spectators present who could have been injured if the driver lost control; 'What a dumb f…! How stupid can you get. He could have killed everyone there but no "cops are good drivers, we can handle anything".' Others maintained that it was on private property and obviously somewhere isolated 'there is no law against doing that on private property, … if anything you could say that he was encourageing the boys to keep it of the road.' (http://au.youtube.com/watch?v=_fGOx7o17RM accessed 27.11.07, withdrawn soon after.)

The police represent both the expression and repression of rebellious desires such as the desire for thrill and speed in cars. Police are not a straight forward emblem or agent of governance and have thus had a mixed relationship to the governance of motoring. They are a significant part of the governance process and can use their power to suppress and deny or allow displays of driving skill as they see fit. Young drivers are more likely to receive attention from the police than other drivers who nevertheless consistently drive in irresponsible ways. More casualties result from drink driving and speeding amongst the general driving community than from illegal street racing and yet more attention is given in the media and by police to the latter. While street racing cannot be condoned or left alone there is a greater willingness to scapegoat the young than to confront the overall problems created by the car in combination with values of combustion masculinity, power and individualism. The desire for the car and for speed and power promoted through advertising and central to consumer culture and the increasing value placed on constant excitement and stimulation are the problem. The police also embody these values.

Governmentality and self-control

Governance generally involves control exerted from outside and enforced through the authority of the law. A form of control that extends beyond external enforcement known as 'governmentality' was signalled by Foucault (1991). It includes specialist knowledges and the shaping of particular subjectivities through the inculcation of norms and values and is part of what is considered a trend toward reducing external regulation and increasing emphasis on individual self-control (Reith 2007). The issue of control with cars and driving is quite complex however and governance as well as governmentality both have a role.

The problem is not so much whether control is exerted from outside or inside individuals but the forms that control takes. External control has obvious limits and requires having the authorities consistently visible but it does have some impact and enforces particular values related to driving. If it is exerted inconsistently however, it has more limited impact and if some of the values it enforces are conflicting then the conflict has to be masked in some way.

The media plays a part in the construction of subjectivities suitable for the culture of consumption that pervades western societies by helping to frame objects as well as subjectivities (Couldry 2000a). Manufacturers appeal to and reinforce particular values through the media as well as through their products. For example by producing and promoting cars with speedos which now go up to 240km/h in standard cars and even higher in sports cars, the ongoing expectation of higher speeds is continually reinforced.

In many areas control exerted through external things and conditions is emphasised in technologised societies at the expense of interpersonal and personal control over one's life experiences. The focus on external forms of control is limited and limiting since we can never control the environment to the extent that we may wish. No matter how good we make road networks and cars there is a limit to how much the system can absorb. At the same time control since Foucault has been considered as increasingly internalised so that citizens govern themselves more and more rather than having to be governed by external authorities. Road safety is a classic example of the requirement of self-discipline in Foucault's sense, and perhaps one of the original reasons for the rise of self-discipline. There is nevertheless an ongoing and constant need for law enforcement on the roads primarily because mobile subjectivities are constituted both through discourses of safety and discourses of risk, and these appear to be 'natural'.

Elias (1995) has argued that there has been a continual decline in aggression on the roads because the civilising process that has accompanied the car has brought about a concomitant self-control and this has made drivers more measured in their judgments. Elias also acknowledges however, that there are still too many deaths and injuries from cars. While the civilising process produces forms of self-control that limit violence, the complementary decivilising process proposed by Elias produces articulations that emphasise and glorify being out of control. The play between control and lack of control, often evident in car advertising and discourse related to cars, is indicative of the struggle between ideas of freedom as lack of restriction and the freedoms the car allows (Redshaw 2007). Elias appears to allude to underlying forces of civilisation and decivilisation rather than considering the values identified as expressing them to be part of the culture of particular societies.

The battle for control is generally considered to be a war waged between individuals and the authorities. How individuals and groups relate to transport and car networks, which involve rules that all must abide by is the central concern. Individuals having the freedom to get about in cars with complete independence posed particular social problems and the need for regulation of some kind because of the sheer magnitude of the damage cars can do. The regulation has not been primarily aimed at government overpowering individuals however and has involved the cultivation of particular types of individuality suited to mobile consumer societies. A set of external conditions is required for mobility to be possible at all and increasing regulation is required by the escalation in the numbers of cars and people and increases in the speed of travel (Bohm, et al. 2006, 11). The burden of responsibility has indeed been shifted to individuals but this responsibility does not also cultivate social responsibility in the forms that it has at present. Cars and pedestrians have been increasingly separated, with the car given priority, mainly because it can be used effectively to assert authority and does so much damage.

The character of driving practice has been explored by theorists who consider road safety to be a discourse of social control while car enthusiasm/speed thrill is seen as the expression of freedom, the antinomy of control. Peter Rothe (1994) explores in phenomenological terms the implications for him of the presence of speed cameras. He feels his privacy is invaded and he is made more conscious of himself as being watched. Rothe shows the ways in which cameras and other forms of surveillance could contribute to the internalisation of discipline so that the driver is watching themselves in place of the police. While he acknowledges the value of these forms of surveillance he maintains that they should not be relied upon in isolation and should also include engineering, education and media (Rothe 2002, 312). Road safety then is seen as supporting social control in creating the 'safe' driver who monitors themselves (Packer 2003). Enthusiasts and thrill seekers who oppose and defy road safety messages, their views expressed in car magazines such as *Wheels* where speed limits are often lamented, are considered the rebels who refuse the discourse of safety and assert their 'independence' through noncompliance. Romanticised in road movies and stories, they are seen as the true embodiment of autonomy avoiding the passivity of compliance to safety.

Self-control is a result of 'civilizing and technicizing' influences according to Mike Michaels (2001). Increases in self-monitoring are required by external controls such as traffic lights. Michaels says this is not unidirectional and also notes a process of decivilisation that occurs. Loss of self-control is also in part the result of increased technicisation: '… the car itself is instrumental in this moment of decivilization: it is an actor that, in that it embodies multiple scripts, facilitates rage' (1998, 130). For Michaels the car and road networks have brought about a loss of self-control. Self-control in this sense is a public or social function where people feel a need to insert themselves into an existing dynamic and find ways to control their relation to how it operates. For Michaels the increasing technology of the car and road networks take control away from the driver thus reducing self-control. By self-control then he means controlled by oneself or autonomous rather than strategies of controlling the self.

The self-control engendered by the car is an example of governmentality or the internalisation of disciplinary measures. Discipline is part of what is involved in being shaped by automobility and within the context of a highly mobile society with particular cultural forms. For many however discipline is not merely the shaping of individuals within a social context but the repression of natural individual desires and impulses. It is being controlled by others in ways that we do not wish to be controlled. And yet, as Bohm et al. (2006) point out, although autonomy is a central value in automobility the combination of autonomy and mobility is 'ultimately impossible *in its own terms*'; 'Cars need roads, traffic rules, oil, planning regulations and the representation of car driving as autonomous movement involves disguising such conditions.' (11)

There is an evident commitment to cooperation that makes the road network relatively successful at getting us where we want to go unscathed, and people do internalise the rules of the road as well as the consequences of getting caught doing the wrong thing. Enormous amounts of self-control are required in driving however as there are many contradictions in the cultures of driving that operate within the road networks making such an effort necessary.

Many sacrifices are made in other ways to achieve smooth functioning systems that are relatively crash free and these have been alluded to throughout the book. Quality of life has been reduced by the demands of motorised traffic with the exclusion of pedestrians and other forms of transport from many areas and inadequate development of other modes of transport resulting in increasingly dense traffic. Furthermore the car itself and the road network has required the development of particular types of subjectivity and these have been increasingly moulded by developments in car and road technologies. The emphasis on internal comfort has meant the car is built for the driver and not those outside the car for example. Efficient movement has been the priority of road networks and this has reinforced expectations of unencumbered freedom of movement for motorists. As Bonham (2006) notes: 'Streets and street users were increasingly brought under scrutiny for their potential to facilitate or impede movement.' (61)

With motor vehicles being regarded as the most efficient means of transport, cars were given priority and pedestrians and others faced more impediments (Bonham 2006, 62ff):

> The claims pedestrians made on public space could be readily ignored when rationalizing travel in terms of the efficient conduct of the journey. Those who did not move quickly or who used street space for activities other than travel were themselves guilty of wasteful and inefficient conduct. (Bonham 2006, 64)

Figure 5.6   Traffic order in Barcelona – waiting at a city roundabout

Bonham argues that since some modifications to the efficiency of the system were made to avoid death and injury, safety became the discourse within which 'the performance of the journey and the body of the traveller were all scrutinised' (64). It has been argued here however that the discourse of risk and the performances of driving shaped in relation to gender, age and other characteristics have also had a dominant place in shaping mobile subjectivities. Efficiencies have not tended to be sacrificed even if compromised in high pedestrian areas. Even quiet suburban streets are still the province of cars not children, pedestrians or bicycles. Both speed and safety make up ideas of efficiency of travel as there is little point in having a fast network if the chances of being killed or injured are high. Nevertheless a level of death and maiming is accepted in the name of speed and efficiency.

Self-control has thus had a role in mediating the differing demands of speed, risk, safety and efficiency. Different understandings of personal and self-control operate within the context of the social practice of driving however. On the one hand personal control is the right to control one's movements and actions while self-control suggests the demand to hold in check natural impulses. Control of an external object such as the car, needs to be constantly reinforced and magnified in an attempt to gain a real sense of personal control from it. The car gives the impression of the driver being in control and at the same time exerts control over the driver. Drivers have been created within discourses and practices of consumption and risk requiring new and better forms of excitement and stimulation, as well as safety discourses that are attempts to introduce control of the impulses created within the former discourses. Mobile subjectivities are created within conflicting discourses that appear natural and to follow from one another.

While the car has engendered a form of self-discipline this has not included an emphasis on ownership or social responsibility, in the sense of agreement with regulation as working to the benefit rather than the frustration of individual goals. Road regulation is popularly regarded as coercive and frustrating, and understood through an authority focus where the government is seen as limiting freedom, when it is also enabling. The sense of ownership is then based on resistance to authority protecting the driver's right to drive to his/her abilities and desires and not the rules of the road.

The problem with the way in which regulation is determined and experienced is not that it is based on self-discipline or internalised self-control, but that the self-discipline required by driving a car on public roads is articulated through and tempered by discourses of thrill, excitement and risk derived from rebelliousness. Discipline and punishment is still a primary mode of governing in relation to cars and mobility. The authority figure wagging the finger at the disobedient driver creating the dynamic of outlaw and oppressor/persecutor is the model of discipline potentially internalised as self-discipline, contributing to both the vulnerability and eager aggressiveness of young drivers because of the dynamic set up between authorities and drivers. Meanwhile consumption discourses and market forces create desires for disobedience, the expression of supposedly natural impulses at the expense of others and values of aggressive competition.

The individual driver is ultimately the agent of responsibility in driving regulatory discourses and not manufacturers, but the agency of the man in control is pitched against the authorities who seek to tame the desire produced in promotional discourses. This desire appears as natural but it is produced as a powerful spurting desire that has to be continually repressed by the social context and thus requires an outlet for which the car is the ideal device. The subject of mobility is produced through both a discourse of risk and a discourse of safety. It is not because the car is a dangerous device however that the discourse of safety exists but because the driver poses a risk to anyone. Consequently the regulatory discourses of driving invoke authority rather than self-control which apparently cannot be sufficiently relied upon in the context of driving within these discourses. The idea of the natural desire to go fast and to discount anyone else beside oneself creates the need for constricting discourses of punishment.

The 'Blood on the streets' television campaign from Queensland Transport was shown in 2006 before holiday periods. It shows a red stream spreading through the streets, which are illustrated by simple lines on a white background. The ad is fairly abstract in that it does not show any actual streets, cars or traffic. The voiceover declares in a deep male tone: 'When it comes to speeding the word on the street these holidays is zero tolerance. These holidays the police are out in force with more speed detectors and the full authority to hand out heavy fines and heavy penalties. Why? This year in Queensland speed has already taken too many lives. As far as we are concerned, there's been enough blood spilled on our streets'. The tagline on the final screen reads, 'Every k over is a killer'.

The ad is clearly intended to appeal to drivers through the deep voice of authority but it addresses an anonymous driver, leaving open the space of culpability, though this is clearly drivers and not car manufacturers. Most drivers will happily exempt themselves from the vacant position of driver in the ad and fill that position with the 'hoons' who do 170km/h in 60km/h zones. Emphasising the 'unnecessary' carnage on the road, the ad allows drivers on the whole to consider that they are not at fault or responsible for it. The threat of authority does not produce any further understanding or specific knowledge of why speeding matters or acknowledgement of the need for cooperation.

The deep male voice threatening heavy fines and penalties is at the same time the voice of the driver out to defeat the authorities, and maintains the separation and ongoing battle between the free driver and the authorities that try to regulate them. The intention of the message might be to appeal to drivers as a community, however it fails in drawing on the traditional distinction between the free individual and the strong social forces attempting to tame them against their will.

The social and cultural contexts that such campaigns are entering into involve some deeply maintained ideas of control that are opposed to the control imposed by authorities. Firstly, the driver is seen as in control, in having control of the car and driving it according to his ability. The car and the road reportedly offer the temptation to speed making the driver innocent of culpability when he loses control and 'inadvertently' disobeys the rules. Drivers maintain their sense of control, and loss of control is popularly considered to be due to lack of skill or mastery, not poor judgement or a wrong decision.

The battle for control is a war waged between red-blooded males teamed up with their cars, and government authorities. The male desire to show off and demonstrate ability to control the car is seen as a 'natural' wildness and aggressiveness that has to be tamed by the strong arm of the law. The car itself remains innocent but when combined with uncontrollable, unstoppable male desire or 'combustion masculinity' it makes government control necessary. Road safety campaigns themselves appear to maintain this idea of masculinity and control in that they make no attempt to emphasise self-control on the part of the driver or to consider the dangers of cars resulting from the ways in which they are promoted as well as their weight and force.

In the research on road crashes problem drivers are described as having a tendency to overestimate their level of skill and control and maintain overconfidence and unrealistic optimism in their driving (Gregersen 1996; Harre, Forest and O'Neill 2005). On the other hand problem drivers are considered as having skill deficits or uncontrollable impulses (women and young drivers), as disobedient or nonconforming and therefore having to be punished and made to submit to authority and regulation or as risk-taking (young men) and therefore having to be excluded from the driving population. Problem drivers are only disciplined however if they are caught and prosecuted, and the promotional discourses produce a driving subject that is prepared to take risks and to seek excitement through speed and challenging driving so that drivers who break the law and get away with it can regard themselves as successful subjects of mobility. Getting caught is often considered bad luck inducing regulatory campaigns that state, 'Drink driving is not bad luck it's a crime.'

Speeding has continued to escape the serious stigma of criminality because it is regarded as acceptable within some limits as to how much speeding and where it is engaged in. High range speeding up to 200km/h on straight stretches of open country roads, particularly in remote areas of Australia, is acceptable whereas doing 10–15km/h over the limit is acceptable in urban areas. Young drivers often fail to make the right distinctions as to when breaking the law is appropriate. Women on the other hand have tended to be cautious drivers and this has resulted in men seeing them as incompetent. Younger women are more prepared to speed within the socially deemed limits and to drive aggressively engaging in close following and accepting the socially prescribed necessity of packing as much into every hour of every day as possible requiring impatient drives further afield than they otherwise would have.

The 'enough is enough' website offers an opportunity for members of the public to indicate their support for the message, 'enough is enough'. The section on the screen reads: 'Register your support here: 3290 people agree enough is enough. By selecting this link you will show your individual support for the Queensland Government's new road safety initiatives.' However the ambivalence of this support is shown in the next part of the screen. Further down the screen a road safety poll asks: Do you ever intentionally drive over the speed limit?' It is then reported that 53% of respondents have voted 'Yes' and 47% 'No'. Of the more than 3290 people who have accessed the site, 53% state that they intentionally drive over the speed limit.

A number of features of the campaign are worthy of note here. Firstly, by pitching the strong arm of the law as the reason to obey the law, road safety campaigns maintain the idea of uncontrolled male energy requiring regulation. Alternatively or additionally, campaigns could draw on an acceptance of the law because it protects us not because the police have the authority to impose it. Secondly, the battle is often represented as a war with individuals and appeals to lone drivers who appear in isolation from others rather than expressing the social environment of the road. The campaigns are intended for every driver who takes a little bit of a risk, however it is pitched/presented in the context of television where car ads often show a man alone in a car on a country road devoid of other cars – the fantasy of the open road. Add to this the anonymous driver of campaigns such as the 'Blood on the streets' campaign and the tendency to consider our own perspectives as drivers at the expense of others and the context of the ad makes it about a lone individual. There is a threat of grave consequences for the individual of not obeying the law rather than an appeal to the social cooperation of driving in which others can be taken into account instead of closing ourselves off.

Number crunching governance

One of the major critiques of governmentality has been the use of epidemiological research on a large scale to determine management of populations. 'Statistical determinism' uses quantification and numeracy in the planning and operation of 'the various socio-spatial practices included in the management and performance of road traffic and automobility' (Forstorp 2006, 95). The impersonality derived from the abstract objectivity of such research is regarded as a favourable way to develop appropriate measures to counteract the problems created by the prevalence and freedom of motor vehicles. Objectivity can to some extent overcome the privileging of some forms of mobility at the expense of others and this has finally started to happen as pedestrian costs become more apparent, although these can still be hidden in a plethora of statistics. In looking at available global statistics it is difficult to gain numbers on pedestrian or cycling fatalities in China for example where the emphasis in China's presentation to the United Nations Economic and Social Commission for Asia and the Pacific (Available at http://www.unescap.org/ ttdw/roadsafety/countryreports.html) is on building beautiful roads for the growing numbers of motor vehicles invading the traditional streets every day. India's report on the other hand focused more on recognising different traffic needs although it did not give specific statistics on pedestrian and cyclist fatalities. The World Report on Road Traffic Injury Prevention (WHO 2004, 42) gives a comparison of road users killed in various modes of transport as a proportion of all road traffic deaths showing the proportion of pedestrians killed in Delhi in India as over 40% of the total and cyclists more than 10%. Overall it says pedestrian fatalities account for between 41% and 75% of all fatalities (41).

Recognition of the dominance of motor vehicles and correcting the imbalance has been a long time coming despite such objective research. If the basic premises are not examined and critiqued regularly then biases will continue. There is often little connection with the social, political and cultural context of the research and little willingness to challenge dominant political values hidden behind the apparent good intent of road safety. Since it is the rich in professional positions of influence who are gaining increasing access to cars in countries like China and India there will be more interest in protecting and supporting the desires of car owners over those of the poor who are the most vulnerable as pedestrians, cyclists and two wheeled vehicle operators. Motor vehicle manufacturers support and promote the interests of drivers without any concern for the likely victims of fast and powerful motor vehicles:

> In apportioning blame for accidents, users of either soft or hard means of mobility are treated as equally culpable, when it is those using harder means who have a potentially destructive weapon at their disposal. (Freund and Martin 2002, 110)

The use of statistics for social planning incorporates discursive and ideological formations and planning imperatives, that is, they are 'explicitly ideologically and policy oriented' (Forstorp 2006, 97). Several paradoxes are encountered in the rationalising process:

> … speed is associated not only with enjoyment, autonomy and freedom, but also with accidents, crime, sin, violence, and other 'major problems'. Governing the politics of speed includes coping with subject's interpretations of times, spaces and desires, while simultaneously promoting speed through the development of safe automobiles, 'tolerant' infrastructures and 'forgiving' transport systems. (100)

Figure 5.7   Provision for cyclists and people – a cycling lane as well as a retreat in the middle of a busy Barcelona street make space available to uses other than motoring

The promotion of speed has occurred on two fronts, as has been argued here; that of the motor vehicle industry and lobby groups and other industries that demand time savings that can be turned into increased profits, as well as that of the regulatory authorities who have shaped the road networks to cater to the speed and efficiency of motor vehicles at the expense of other possibilities for mobility. Consumers as drivers have also made demands on the system that have been catered to by governments keen to address efficiency.

The extent to which we have become car centred is beyond individual desires and demands however, and has to be considered on a social and cultural level. As cultural theorists such as Williams and Morse have shown the car is connected to other changes and developments within industrial and postindustrial society such as the separation of work and home, and the advent of television and shopping malls. Ideological perspectives have contributed to fostering the dominance of the private motor vehicle, particularly *laissez-faire* liberalism (O'Connell 1998) and more recently, neo liberalism (MacGregor 2002, Forstorp 2006). Societies where market forces influence much of the social world vary in policy from those where government mediates commodity relations to a greater extent such as Sweden (MacGregor 2002, 127–128).

The individual is the focus of attention on every front from the target of advertising to the discipline of authorities when there are many aspects of the whole system that need to be addressed and not just individual personalities, desires, attitudes or behaviours. The system as it is now promotes desires that must be tempered by the regulatory system, which fundamentally supports those desires as inevitable. Confronting desires as created and channelled within the context of social and cultural demands means that they can be questioned and assessed for their value rather than being treated as necessary. Desires in postindustrial societies are channelled towards continued consumption and the shaping of identities through consumption. Making greater demands of other aspects of the system rather than merely focusing on individuals who vary from a norm that is established by large numerical studies is much more equitable and likely to produce a more responsive system overall.

The desire for recognition is also shaped in significant ways in relation to cars, especially for men, and particular emotions are fostered within the context of driving when a range of emotions are significant in the engagement of people with a practice such as driving. As Mimi Sheller states:

A better understanding of the cultural and emotional constituents of personal, familial, regional, national and transnational patterns of automobility can contribute to future research programmes and policy initiatives that resist the powerful yet ultimately unsatisfying aggregation of social data based on statistical quantification of individual preferences, attitudes and actions. (2005, 222)

Recognising emotional engagement as significant and important rather than abhorrent and investigating different types of engagement as well as encompassing individual differences in mobility needs and engagements would mean better capacity to address mobility issues. Rational explanations must extend from emotional engagement rather than deny it. Showing alternative connections to the excitement/aggression connection could frame new possibilities for being within driving cultures. Similarly emphasising rather than absorbing the inconveniences of the car would help to promote positive valuing of alternative forms of mobility and recognition of the source of some of the frustrations and limits of automobility as within the system that continually demands ever increasing speed as well as convenience.

Chapter 6

An Ethical Future of Mobility

In this final chapter I would like to conclude by emphasising the importance of addressing the social and cultural issues and aspects of mobility and of providing potential images or visions of the future of mobility. Bleak images of the future abound in movies like *Mad Max* (1979) and *The Fifth Element* (1997) but it is also possible that transitions could be made to new technologies and ways of being with mobility with a growth in the positive aspects of existing social and cultural worlds. Firstly there follows some discussion of the alternative futures for mobility that have been outlined by John Urry for example and some that are already in use in different parts of the world. The need for speed and alternative masculinities are considered followed by different visions of mobility through planning, ideas of hospitality and campaign visions.

Technology is often heralded as the solution to reducing road death and injury rates, increasingly removing the human factor from the equation. However the values and social norms underlying the current dominant form of mobility cannot be ignored in confronting mobility issues into the future. In current understandings of mobility the driver or operator of a motor vehicle is seen as the one in control, and the passenger merely subject to that control. An agent of mobility however is not just a driver or operator of a motor vehicle but is at times a pedestrian, a passenger, and possibly a cyclist. The boundaries between driver and car have become increasingly blurred with new technologies but remain significantly distinguished in cultural terms. Motor vehicle manufacturers emphasise the value of control in advertising their new models even when they have added greater safeguards such as stability control. Considering individuals as agents of mobility avoids posing an opposition between the driver or the vehicle as the agent of control based on current emphases on driver centred values, particularly as speed controls are likely to be applied in the future. Pedestrians and cyclists as well as bus and train commuters or passengers are then also agents of mobility thus avoiding the idea that only drivers of motor vehicles are in control of their mobility.

The ways in which individuals relate through different forms of mobility, whether public, private or variations on these, is another area where values and social norms are important. The emphasis on the value of privacy in cars operates to obscure the fact that mobility is essentially both private and public and above all social. How these are brought together and resolved is central to future 'safety' and well being, as well as accessibility and the possibilities that can be imagined, of mobility. A move from the individual focus of marketing to a social/community focus is thus central to imagining mobility into the future. This requires a move from car and driver centred to a greater recognition and focus on the social and community aspects of mobility so that places are designed primarily for people and not for motor vehicles.

Shifts in values are already apparent in the increasing emphasis on the need for changes in the human impact on the environment and the value placed on the quality of the human environment. Planning has arguably centred on the most able aggressive users rather than the most vulnerable users however the tide is changing in this area. Designer Jan Gehl primarily designs cities for people, that is, cities are designed around making them pleasant places for people to be. Pedestrians are given greater priority and public spaces for people to gather and linger, are central. Using the theme of hospitality I will outline some aspects of mobility as they currently exist that could be expanded into the future.

Alternative patterns of mobility

John Urry (2005) has outlined potential post-car patterns of mobility indicated by six technical-economic, policy and social transformations that in their 'dynamic interdependence' might bring about new systems of mobility. These include not only new forms of fuel from electric to hydrogen and methanol fuels, and new materials for car construction making cars lighter and needing less powerful engines, but also 'smart-card' technology. A central feature of these innovations is the loosening of the bind between drivers and privately owned vehicles as well as the attachment to the internal combustion engine and its 'fire power' (Virilio 1986).

The information transfer involved in the 'smart-card' referred to by Urry (2005) could help to de-personalise cars making them more like 'portals' through a single means of paying for transport and adapting it to one's own needs. The card would hold information on one's own 'settings' for seat, mirrors, entertainment and so on, and simply inserted, all 'settings' within the vehicle are adjusted automatically thus overcoming the problem of having to adapt a vehicle to one's own needs one at a time. This is likely to be effective in large city centres where mobility through private cars has become increasingly difficult and people are prepared to accept other ways of being mobile and are less likely to own their own cars. Cars are also being de-privatised through car sharing, car clubs and car-hire schemes, particularly in Europe.

Paris is trialing in 2007/2008 a scheme of bicycle 'sharing'. Bicycles are housed at various locations around the city and can be employed at one point and left at another using a card with information about the user. Police in Paris also use bicycles to facilitate mobility through the city. Bicycles are not as common in Paris as they are in other cities such as Amsterdam and Barcelona is the city of the motorcycle. Other cities in Holland make great use of the bicycle in all weather conditions. It is hoped that the bicycle program will result in more people being prepared to give up their motor vehicles in favour of a friendlier form of mobility that is not only less polluting but takes up less space on roads and is less violent.

The recognition in transport policy that merely providing more roads on the basis of predicted usage only increases car dependence is noted by Urry (2005), as are alternatives being developed, including computer-mediated intermodality, integrated public transport, cycling and pedestrian facilities, and advanced traffic management. Meanwhile, communications including the internet, are creating 'hybrid mobilities' and potentially reducing the need for travel. Urry optimistically entertains the possibility of a tip in consumer demand and other factors that will create new systems of mobility. While attachment to the internal combustion engine powered private car is going to take some time to erode, it is important that alternative possibilities are being imagined and developed. Not only are those outlined by Urry being entertained, they exist in meaningful forms that are already having an impact on the way mobility is being thought about. A new system is likely to emerge in unpredictable ways nevertheless, Urry warns.

Figure 6.1   Bicycle borrowing scheme in Paris

Alternatives being imagined need to include a sense of sociality and not just an overwhelming emphasis on individual privacy. Privacy is often imagined as in opposition to undesirable company represented in news media as ready to invade and steal from the innocent. To a large extent mobility is social and only possible because it is social. We share the roads even from our individual private vehicles and we cannot have our individual mobility without a shared system – whatever forms mobility takes it requires access to public spaces dedicated to mobility. Autonomy as an ideal of mobility is impossible as Bohm et al. (2006) have argued, and must also be disentangled from mobility in the ideal of automobility. Complete autonomy of mobility is a misnomer misleadingly applied to automobiles reinforcing the impression of singular unimpeded activity. While individuality of mobility is possible in the sense of deciding where one wants to go and when and to varying degrees how, these choices are all made within the limits of what the system allows as are all consumption possibilities. The illusion of free choice allows consumers to think the choices are theirs for the making while it is well known that producers and retailers use 'tactics' to get us to choose their products and only produce within a limited range specified by other producers. We are not free to use light cars that run on air except in principle since no such vehicle exists and oil companies make it very difficult for any other kind of vehicle to be viable. Consumer demand can have an impact but it can mean initially choosing inconvenience.

Another emerging pattern of mobility is the use of minibuses for various purposes. Small buses are used by pubs and clubs to transport people home locally if they have had too much to drink and are unable to drive themselves or have previously

Figure 6.2   Community transport bus for older people

arranged being driven in the bus. They are also used to transport older people who are no longer able to transport themselves to day centres and appointments. Youth buses have been trialled in areas on Friday and Saturday nights where public transport is scarce and young people are forced to drive in order to go out. Hotels often run small bus services to airports and tourist services that are catering to smaller niche groups also use these vehicles. These 'community' services are likely to become more common even though there are problems with them such as the difficulty of frail aged people getting on and off the buses and the behaviour of young people on such services. Neither individual nor mass transport, such services are both public and private and require behaviour that is perhaps more 'familiar' than mass transport but certainly more oriented to the presence of others than individual private cars.

Addressing the apparent need for speed

What do you feel?

I feel the need for speed! (*Top Gun*)

The 'need for speed' in contrast to the need for mobility, has dominated the development of motor vehicles and road transport systems. Even though it is aggressive and promotes dominance and in current technologies optimum speeds have already been reached and speeds are being lowered for liveable communities there continues to be an emphasis on the need for speed. It is possible in programs such as Sweden's Vision Zero to lower speeds and use other measures related to road and vehicle design to drastically reduce death and serious injury on the roads however the pro speed lobby is made up of powerful forces. Roads have been designed to foster the expectation of speed as has been shown throughout the book.

A common sign on Australian roads when there are two lanes in each direction is 'keep left unless overtaking' but this sign has the effect of favouring speed at the expense of local and slower traffic. These signs can be found in New South Wales even in highly built up areas where there are right and left turns onto and off highways and they tend to reinforce an expectation that through traffic will not be inconvenienced by slower or local traffic. It is common for the traffic in these lanes to be travelling at more than the speed limit and for there to be significant demand for those only travelling at the speed limit to get out of the lane. This can make turning on to and off these roads very difficult and sometimes dangerous. When wishing to turn right one has to move over into the right hand lane in front of traffic wishing to speed through and can often be bullied through close following or flashing of lights for doing so.

Some groups in Australia as in other countries argue that speed limits should be determined by what motorists are comfortable with or by the 85 percentile of the speed of the traffic. *Wheels Magazine* (John Cadogen 2004) on their 'Safety Debate' page claimed speed was what facilitated transportation and that whether speeding was dangerous or not had as much to do with the 'appropriateness of the posted speed limit' and further that speeding 'is not always unacceptably risky, as the thousands of drivers who speed every day can attest' (29). It is stated that for

**Figure 6.3   Keep left unless overtaking on a busy highway within kilometres
of homes and shopping centre as well as numerous side roads
entering**

physicists speed is just 'the rate of change of distance with time', but this would
only be the case in an abstract world. Actually movement is what gets us from A to
B while speed is the rate or rapidity of movement and a chosen speed might be fine
for a particular driver but not good at all for anyone else around them.

This view gives complete priority to the motor vehicle driver at the expense of
other agents of mobility, particularly non-motorised road users, and at the expense of
quality of life. Most would agree that speed of motor vehicles cannot be individually
determined but it is more difficult for drivers to accept that speed must be determined
socially for the benefit of the community, not just drivers. High speed in motor vehicles
above current limits is unsustainable not due to lack of skill as is often claimed, but to
the inherent dangers and limits of internal combustion engine powered motor vehicles,
the roads, weather conditions and growing numbers of vehicles.

Many drivers who regard themselves as highly skilled do not respond sensibly in
conditions of low visibility for example continuing to travel at high speeds regardless.
Skillful drivers can be seen daily continuing to speed in high traffic conditions on
motorways, chopping in and out of lanes to sustain their speed and then having to
pull up abruptly and at great risk to everyone when the traffic stops entirely. This
racing, skill based and excitement packed style of driving so inappropriate to the
roads is actively marketed by manufacturers in their advertising and follows from
the computer and video car racing games which attempt to be as 'real' as possible
even to the extent of paying royalties to car companies to use their models in the
games.

Nissan have recently produced the 'fastest and most powerful car to come out of Japan', a 'true hero car', the GT-R (Spinks, Dowling, Hudson and Blackburn 2007). The car has a claimed top speed of 310km/h and a speedo up to 340km/h, the engine 'sounds docile' ambling along at 60km/h with 'no hint of the terror about to be unleashed'. The car is familiar to millions of Gran Tourismo players around the world who have already 'driven the new model in virtual reality'. It is the typical car desired by young men and much more affordable than a Porsche though at $150,000 (AUS) not too many will be purchasing it until it comes onto the second hand market. Nissan hope that by getting 16 year old kids hooked on the GT-R they will feel good about the Nissan brand and buy a Tiida. But will they drive it as if it was a GT-R? Nissan realises the wide recognition of the model is due to the game and even provided Gran Tourismo's developers with the 'GT-R's top secret computer blueprints months before the car was released'. The lines between reality and fantasy are blurred in the overlap between game and car and the game certainly teaches a particular type of fast and aggressive driving style. Motoring organisation NRMA recognises the benefit of simulators as a teaching device but they have to combat the style of driving adopted from simulation games such as Gran Tourismo.

The central emphasis on speed and power in the development of cars such as the Nissan GT-R is essential to maintaining the image of the car as an aggressive beast to be wrestled with by the able (male) driver given the competition it already faces in the market place. According to the review the GT-R is unashamedly driver centred in its aim to make the driver feel at one with the car; 'It is immediately obvious the driver is a priority in this car.' There is no consideration of the context of the roads in the dialogue related to such a car. The car stands alone ready to take on the skill of the driver who wishes to challenge himself, although it is also 'idiot proof' because 'the AWD system calculates the ideal front-rear power distribution by continuously monitoring speed, steering angle, car angle (yaw rate) and G-force'. Dubbed 'Godzilla' the car is shown on a race track with the backdrop whizzing by in a blur.

The Nissan GT-R is a good example of the contradictions of the car. The cherished speed and power are nowhere applicable on the roads and the image demands a driving style that is hostile to the lived environment. Too much emphasis on the driver and the added power and speed that such a car gives him has reduced the hospitability of the roads and allowed the acceptance of road carnage as the price of efficient mobility. The blurred lines between fantasy and reality are central to the value of such a car since its capabilities are beyond the conditions of the roads and the car itself is out of reach of most people except through a simulation game that is widely available.

On the Champs Elysée in Paris Toyota have a showroom displaying two prototype cars. One is displayed in the corner window looking out on to the street and there are typically men of various ages looking at it since it is the car of everyman's dream, a racy sports model. Inside is Toyota's answer to the Smart Car found in increasing numbers all over Europe. This little concept car is likely to go into production in the future whereas the sports model is not because it is too expensive and contradicts demands for vehicle manufacturers to contribute to environmental sustainability. The image remains nevertheless associating Toyota with the tough racy power and speed of the sports car of the future in a world where the environment is of little consequence. Manufacturers sustain the values of power and speed, competition and aggression as central to the driving experience.

There has been greater demand for smaller cars in recent years and the rise of some popular small car models like Smart, the reappearance of the Mini and Volkswagon and in Australia favouring of models like Toyota's Corolla over the traditional Ford Falcon (Spinks et al. 2008). Despite this trend the manufacturers have continued to cater to those who are interested in nothing other than speed and power and can pay for it. Audi, Saab and BMW compete over which has the most powerful sedan and produce gas guzzling V8s and/or turbo charged engines with absurd take off times that are not necessary anywhere except in James Bond movies and even significantly dangerous on the streets. Lexus and Jaguar are producing V8s for the high end of the market.

For Paul Virilio, theorist of speed, speed is the vehicle of violence; 'The progress of speed is nothing other than the unleashing of violence, or … the means to sustain violence, indeed render it unlimited.' (Virilio 2005, 45) Speed for Virilio is responsible for a perceptual blindness that results from removal from the immediate environment through speed and 'the instantaneous redefinition of the image and spatiotemporal dimensions of the territory passed through' (152). Speed influences the way we see, what we see, the way we live. Continually looking for time savings

Figure 6.4   Toyota's concept car in Paris likely to go into production in 2008 possibly as a hybrid

has become obsessive to the point of being the only factor that matters. The trucking industry can make savings by spreading out the costs of faster speeds so that the broader social environment pays. A system that allows a loose distribution of costs for gains in profit is not providing the governance required by the violence of speed. Ralph Nader proclaimed 'unsafe at any speed' in 1965 but we have since found ways to improve car design so that the impact on those outside the vehicle is not so severe. There is still a long way to go however with new kinds of vehicles such as SUVs creating new problems and the desire for more speed and aggression continuing to impel the violent dominance of motor vehicles.

Mobility and accessibility planning

Designers such as Jan Gehl place the emphasis on quality of life for people and making it easy for people to gather and to move around. Gehl has worked on Melbourne, London, Zurich and many others including a recently completed report on Sydney, after successfully showing his principles of bringing people back into cities in Copenhagen since the 60s.

> Jan Gehl and GEHL Architects have been advising the city of Melbourne over the last 20 years in a carefully planned and executed process to revitalise a city centre previously notorious for its poor quality of public life and transform it into a people oriented city. The process has resulted in a documented increase in citizen's and visitor's use of all forms of the public realm including pedestrian streets, promenades, public squares, and public parks. (http://www.gehlarchitects.dk/melbourne.asp, accessed 7th January, 2008)

> The City has significantly improved its walking environment through a coordinated program of streetscape improvement works. The physical improvement of the city's streets and lanes provide for the safety, comfort and engagement of the pedestrian, inviting popular use within a wide choice of through-city routes. (Gehl Architects 2004, 1.4, 20)

> With the introduction of a Congestion Charge London is entering a new era where car dominance is to be substituted by a better balance between vehicular traffic, public transport, cycling and pedestrian traffic. 'Towards a fine City for People' describes the present walking conditions in London and pinpoints the types of barriers and obstacles pedestrians have to fight when walking in London. (http://www.gehlarchitects.dk/london. asp, accessed 7th January, 2008)

Increased speeds for motor vehicles are possible within reasonable parameters but are likely to be very costly. Car manufacturers cannot continue to drive demand and extend expectations of higher speed since there are many other factors to take into account and the bullying ability of motor vehicles should not continue to rule and dominate the streets.

In order to address mobility exclusions greater emphasis needs to be placed on making roads more accessible and friendly for nonmotorised users so that public transport centres can be accessed safely and easily, and cycling and walking are more pleasant and possible. Those who do not have cars are less mobile in today's world because the car dominates. Where there is a good and accessible public transport system such as in Hong Kong, New York, London and Paris there is greater mobility for all but this must be combined with pedestrian and bicycle access as well as access for the disabled. In countries like China and India where car numbers are growing drastically, planning should include bicycles and pedestrians and not just allow cars to plow through the middle of traditional streets and take over. Emphasis in road safety can focus on the safety of the car driver at the expense of those outside it when motor vehicles nevertheless have the advantage of violence over all other forms of mobility. Seatbelts protect occupants of motor vehicles but do little for those outside it.

The most vulnerable road users, pedestrians and two wheel motorised and nonmotorised vehicles, represent between 41% and 75% of all fatalities in lower and middle income countries. As is reported in *The Lancet* (Ameratunga, Hijar and Norton 2006) despite reporting difficulties 'most of the estimated 1–2 million people killed on roads in 2002 were not car occupants' (1535).

Not only must motor vehicle manufacturers no longer be able to influence factors such as road speeds, there should be greater demands for changes to the increased emphasis on aggressivity in design of cars as well as monitoring features such as speedometer ranges and demanding more responsible advertising images. Showing traffic scenes rather than open roads with no other signs of life would be a start and removing references to top speeds and acceleration rates, aggression and power would change the social framing of cars in significant ways. Manufacturer's websites

Figure 6.5   The increasingly popular Smart car found in most European cities

require monitoring as these are becoming sites for more explicit advertising when demands for more responsible television advertising get in the way.

In Sydney in early 2008 the motoring organisation NRMA has complained about the money and space being allocated to a bicycle lane alongside a new multimillion dollar motorway. The President claimed the space would be better spent on another lane for the motor traffic and the cost of 7.6 million on 25 cyclists a day was unjustified. The Roads and Traffic Authority is confident more cyclists will use the dedicated facilities (Smith 2008). Many have since pointed out that motor vehicle drivers also rarely foot the bill for the real costs of roads. More significant however was the reported fact that bicycle sales in Australia have increased enormously even overtaking the number of cars bought in a year by 40% and the New South Wales Roads and Traffic Authority reports a 45% increase in bicycle traffic in the Sydney CBD. Sydney Lord Mayor Clover Moore (2008) reports a plan for the future that includes cycleways: 'While there are major recreational cycleways – such as the Sydney Harbour route and the planned Alexandra Canal path – the city's cycle strategy aims to create an effective and accessible network with major routes less than five minutes' cycle from every residence.'

An ethics of the road requires a new vision of the roads to replace the current vision of aggressive competition and combat with cooperation and shared responsibility. A vision is not an ideal that ignores or attempts to replace reality but a new way of imagining and replacing the existing vision. It is often imagined for example that other drivers are out to make our lives difficult and deliberately obstruct us in our progress when most are simply trying to get where they are going like everyone else. Nor is it necessary for drivers to become racing and rally drivers, rather it is important that there is more give in the system on the part of drivers as well as the technology and the roads.

Alternative masculinities

Combustion masculinity has been associated with speed and power as indicators of high performance as discussed in Chapter 3, however there are other forms of masculinity demonstrated through different types of vehicles and manifested in skillful operating of machinery that demands patience and careful, slow handling. From bobcats and forklifts to dozers and cherry pickers, and 'hiabs' on trucks for removing heavy materials, modern machinery requires skill that is patient and precise rather than aggressive and accelerating. Hydraulic machinery shows evidence of the experience of extending the body where bobcat bucket or a crane becomes an arm which can be manipulated as if it were attached to the operator's body. Building materials are frequently delivered on trucks with a single operator who removes the materials using a hydraulic arm or hiab attached to the truck.

Tree loppers use 'cherry pickers' to get to tree branches along the streets and cut them away from power lines. Operators move the car or bucket using controls on board to precisely locate the branch they want to cut and then extend a saw blade attached by a hydraulic cable to the car to remove the branch.

Figure 6.6   Display of hydraulic machinery. Such machinery requires careful and skillful placement. Photo by Leta van der Waal

An articulated Franna crane 'bends' in the middle allowing it to access difficult sites. The boom on the crane extends out in front of the heavy body to a precise point, lifting a heavy steel girder and gently lowering it into place. Bobcats spin on a dime and manage to leave intact buildings and trees located close to where they are being operated. The precision of the driver is such that the machine is almost endowed with the attributes of a biological being, reaching out like a hand to scoop and move soil, dig trenches and remove tree roots and large plants.

A masculinity based on the precision of operating hydraulic machinery is one that is able to move slowly and carefully, taking into account the surroundings and the impact on other things. Hydraulics in this sense operates as a source of mechanical force or control that is not aggressive and in fact requires quite a different constitution from the competitive, aggressive power of combustion masculinity. As a model or type of masculinity, hydraulic masculinity is everywhere but at the same time is less visible due to the current preference for displays of combustion masculinity.

Hospitality on the roads

The roads and road networks are surprisingly forgiving. Young men chopping and changing between lanes in peak hour traffic on freeways at speeds up to and over 100kms/h are able to survive because most drivers stay in one lane with a reasonable distance between cars. The traffic forms a pattern that they are able to exploit and which absorbs their risk-taking as other drivers back off and leave room. Speeding drivers on freeways are largely successful because other drivers will move out of the way and give them the 'fast' lane. One of the problems that arises in these circumstances is that young men come to believe that it is their skill that allows them to survive and succeed with their aggressive and high risk driving styles rather than that the system has allowed them plenty of leeway. It is a good thing that there is room to absorb mistakes but the emphasis on this kind of driving skill is misplaced. Expectations of higher speed and the level of acceptance of speeding as a demonstration of appropriate skill are highly problematic and create enormous problems soaking up much of the good will and malleability built into and created within the system by the majority of drivers.

While aggressivity is arguably the dominant paradigm of the roads and car design the road system nevertheless operates through a high level of hospitality. This may not be evident as aggressive behaviour has become more prominent and thus tends to be noticed more, whether aggressive behaviour has actually increased or not. The sheer volume of vehicles on the roads has increased and every better, faster road that is built is soon clogged with increasing traffic volume. This also tends to reduce evidence of hospitality. When drivers have waited in long queues for considerable amounts of time they become less inclined to let others in, especially if they may have gained some advantage by moving faster in another lane or along another route.

Dominant discourses of risk and skill have maintained that drivers who abide by the rules are not 'good' drivers but merely 'safe' drivers. 'Good' drivers are those who model themselves on racing and hone their handling skills to take corners faster and drive at higher speeds because they 'know what they are doing'. They are in tune with their vehicles and the road and merely have to watch out for the mistakes and annoying habits of others. This style of driving is at the same time in tune with the demand for efficiency, often at the expense of quality of life. These sorts of values need to be questioned and replaced and other values given a chance to emerge more strongly.

Ghassan Hage in his article 'On the Ethics of Pedestrian Crossings or Why 'Mutual Obligation' does not belong in the language of neo-liberal economics' (2000) takes the example of the pedestrian crossing as experienced by a Lebanese migrant to Australia to reflect on social obligation and the meaning of 'mutual obligation' in neo-liberal societies. Ali experiences pedestrian crossings as 'magical' since cars actually stop for him at the crossing and Ali then experiences a 'buzz generated from the moment of recognition' (2000, 29) making him feel 'important' as a human being. Hage refers to the recognition related to the pedestrian crossing as 'offerings' as these incidents are forms of 'social gifts', 'gifts that society offers its members' (29).

While drivers and pedestrians experience and negotiate the crossing in a multitude of ways, underlying and constituting this plurality of modes of interaction according to Hage is 'what constitutes the single most important aspect of the phenomenon: a pedestrian crossing is an ethical construct' (30). The modes drivers offer to 'convey' the gift and the modes chosen by pedestrians to receive it vary, but what remains is the crossing as a 'structurally present ethical space':

> ... a space where people can enact a ritual of stopping and crossing, and through which society affirms itself as a civilised (that is, ethical) one where dominant modes of inhabitance are invited to yield to marginal modes of inhabitance'. (31)

It is only in a society where members are honoured or valued that 'mutual obligation' can exist – where governing bodies feel 'obligated' to offer spaces that 'honour' its members as important human beings and these members in turn feel an ethical obligation towards the society. For Hage this obligation is 'nothing other than becoming practically and affectively committed' on the part of the members of such a society (32). It is therefore a commitment to an ethical structure that works for the benefit of the members of society and not just the dominant modes or those who are most aggressive. It is also a commitment to 'our own sociality and our desire to live communally' (32):

> A society maintains itself as an ethical community by continually offering to its people the very ethical conditions it wants them to be 'obligated' to reproduce. (Hage 2000, 32)

For many, simply stopping at a crossing is a courtesy even more than it is a legal responsibility. Commitment and courtesy are important elements of hospitality and of making rules relevant. As Hage emphasises however, this is a two way process. Recent political thinking has involved the idea that individuals owe society for the privilege of being a participant. The state's obligation is reduced to delivery of services and emptied of ethical elements like honour, recognition, community, sociality and humanity (34).

The idea that by entering a society we immediately have to give something back assumes that the individual has no value to offer, instead of the presence of any individual being regarded as a gift. For Hage regarding the presence of another as a gift is simply a recognition of their humanity. Asking for payment from citizens then is asking them to earn their humanity, negating 'the very humanity their presence brings' (36). Moving to a framework that is more concerned with giving and responding to others rather than considering that they owe us something requires moving beyond the blame game. Society does not owe us anymore than we owe society anything. It is a matter of assuming that we are being offered something and that each individual offers something to society.

The pedestrian crossing is a good example of a shift in thinking and one that is reasonably well established in Australia as a valuable tradition. Hage's interviewee is impressed with the fact that cars stop for him at a pedestrian crossing in Australia unlike his native Lebanon. It makes him feel valued and recognised. Since around 60% of the world's road fatalities are pedestrians this is an enormous area of potential. How did the tradition become so well established in Australia when in other countries it is unthinkable even at crossings where there are traffic lights that the traffic would stop for pedestrians? While not all drivers in Australia happily stop for pedestrians and Australian drivers could be much more aware of people as having a right to occupy streets, particularly urban streets, many willingly stop for pedestrians as a matter of course and even slow down where there are children on or near the street. As increasing numbers of cars populate countries like China and India basic rules of engagement need to be established that include concessions to other road users. It is not always possible to separate motorised and nonmotorised road users so part of operating a motor vehicle has to include giving way to other road users in some areas, particularly urban areas. Noone has more right to mobility and yet this is what is often assumed with the most potentially damaging forms of mobility. The car allows might to be right and requires considerable governance to redress the imbalance.

Hospitality as an ethics for the roads counteracts the aggression that appears to dominate at the moment. Hospitality on the roads points to the give and take involved in the social gift of the roads and this can be emphasised a lot more. It does need to be assisted by approaches to governance that allow more for the opportunity to emphasise giving and recognition of the sociality of the roads. Governance related to the roads is often nothing more than the authority of the law and the threat of punishment seen as repressing 'natural' desires and required to control 'natural' impulses created through the car. The governance shaping drivers in the context of market driven consumer societies includes social framing by car manufacturers through their advertising, associations with car racing and electronic games, and industries that make gains in time savings through higher speeds, as well as the regulatory authority of governments. Governance can be applied in more ethical ways that involve many sectors of the community in developing a future that is more hospitable.

Informalised authority

It will take more than laws and technology to effect real changes in the significant costs of automobility, both environmental and social. Ethics goes beyond legal parameters and can invoke more appropriate response through community centred engagement. The incorporation of informality into public political discourse or the use of 'informalised authority', can help to establish a connection with the public minimising the social distance and power differences and bringing with it greater involvement of the private sector and community (Lazar 2003). Such an initiative has been instigated in Singapore through a 'courtesy campaign' that incorporates nonformal structures and involvement of different sectors of the community. Given that Singapore has one of the lowest rates of fatality per 100,000 of population in the world (WHO 2004, Table A.4, 195) there are some gains to be made by looking at how road safety is managed.

'Informalised authority' is used to establish an 'engaging connection with the public' (205). Rather than addressing the public generally, specific communities were identified and targeted, such as the community of drivers, commuters, school children, retailers and so on. The government itself bowed out of the central running of the campaign and set up a Courtesy Council comprising private sector talent and professionals such as teachers and journalists who organised and coordinated the campaign. Targeted groups then plan and organise their own 'courtesy activities'.

As an example in the transport sector the Automobile Association of Singapore along with the Traffic Police and Land Transport Authority and a number of private organisations have put up banners displaying a courtesy message along public roads all over the island (205–206).

An emphasis on impersonality and objectivity in research is reflected in or reflects the impersonality emphasised in strategies of traffic control on the roads. The strong arm of the law is the final end point of social regulation and control but rather than just relying on the threat of authority as stronger than individuals and able to punish them and take something away from them, appeals could be made on many levels before the authority of the law is invoked. Education campaigns with some kind of vision or overall aim such as Singapore's Courtesy Campaign are more comprehensive and social and with the incorporation of informality and the involvement of community, are potentially more engaging of those such campaigns are most interested in influencing.

Jonasson's (1999) study contrasting the traffic light and roundabout in terms of their facilitation of interaction between drivers provides an example of the impersonality of some approaches to traffic regulation compared with the more personable social interaction of others. 'A traffic culture can … be described as a system or network of actions creating meaning and values produced at a place where humans meet as traffic participants.' (49) At roundabouts there are frequent but not severe accidents. Jonasson argues this is because the roundabout 'forces traffic participants to interpret and consider their own as well as other's actions through the communication of social expressions.' (53) Jonasson describes the interactions that occur at an observed roundabout and the lack of interaction at traffic lights where the lights structure the traffic flow.

Traffic lights in contrast to the roundabout, according to Jonasson, are 'constructed, manufactured and managed in a context where human beings, if considered at all, are viewed as isolated individuals without social relations to others.' (53). The traffic light, according to Jonasson, produces 'artificial hierarchies that aim at being simple, objective, unequivocal and unambiguous' without requiring social knowledge. They can be viewed as an attempt by traffic planners and authorities 'to control and manipulate social relations in everyday space', to over-ride social knowledge which is seen as fallible. An unintended consequence of imposing a different form of regulation is the emergence of 'new hierarchies characterised by competition rather than cooperation.' (53) Traffic lights are intended to regulate behaviour through rational, objective systems but nevertheless maintain a form of emotional engagement concerned with competitive relations. Jonasson's study demonstrates the importance of social interaction and the forms it may take in driving.

While objective traffic regulation is necessary and has positive results, the emphasis on objective control in traffic regulation could be at the expense of social knowledge which needs to be understood and engaged, and which actually influences driver behaviour in more significant ways. In New South Wales holiday periods during which more deaths traditionally occur, a campaign invoking the authority of the police and the law is frequently used. It shows a licence being shredded with the deep male voice of authority stating: 'If you're caught not wearing your seatbelt this … (Easter, Christmas, etc) you'll lose half your licence, and if just one of your passengers isn't wearing theirs that's your licence gone. Double demerits this ...' No doubt the advert has some impact and it at least reminds people that they should be wearing seatbelts, that they can be fined if their passengers are not wearing theirs and that they will lose more points if caught but it sets up the motorists against the authority of the police who 'will get them' if they disobey.

Other campaigns are being used by the New South Wales Roads and Traffic Authority that invoke social judgement such as the 'Country Speeding' campaign (2006–2007) and the 'No one thinks big of you' campaign (2007) or are informative such as the 'micro sleep' ads, however in most campaigns the authority of the law at state government level is invoked to apprehend and control the impulses of the misbehaving driver. The vision in this case remains one of repressive social laws pitched against the 'natural' impulses of car and driver that must be controlled through willpower.

Future visions

Although Singapore's campaigns have possibly had some success this cannot be proven except through continual decline in fatality and injury statistics and that could also be due to other factors. Singapore has a high number of motorcyclist and pillion rider (102 in 2006 and 103 in 2007) and pedestrian deaths (42 in 2006 and 58 in 2007), and pedal cyclists and pillion riders (14 in 2006 and 22 in 2007) compared to motor car drivers and passengers (18 in 2006 and 24 in 2007), and other motor vehicles such as trucks and buses (14 in 2006 and 12 in 2007) indicating that motor vehicle safety is probably adequate from the occupant perspective and highways reasonable for motor vehicles at least for four-wheeled vehicles. Although there are high numbers of motorcycles in Singapore, in 2007, motorcycles constituted 17% (142,337) of the total vehicle population (824,388), vulnerable non-motorised and motorcycle users are possibly not catered for as well as motor vehicles (Statistics courtesy of the Singapore Land Transport Authority www.lta.gov.sg).

The number of speed related deaths in Singapore was reportedly 67 of a total of 190 in 2006 and 70 of a total of 214 in 2007. There is no indication from these statistics as to whether four-wheel motorised vehicles caused or were involved in more of the deaths and this should be considered more closely. At the very least the assumed priority of the car and bullying power of motorised four-wheel vehicles and their impact on other road users and overall quality of life needs to be addressed in the overall vision as does expectations of speed promoted by manufacturers and evident in television and magazine advertising.

Another country that has a consistently low rate per 100,000 and a vision for reducing the road toll is Sweden. Sweden has had a road fatality rate of 5 per 100,000 for 2004 to 2006 (SIKA 2006, 84). In the case of Sweden most of these deaths are car occupants, 261 of a total of 455 in 2006, whereas there were 55 pedestrian and 26 cyclist deaths, while 55 motorcyclists were also killed (23). The Swedish report also gives numbers on the involvement of motor vehicles in fatalities – 98 deaths (21% of total fatalities) involved a single car, and 155 (34% of total fatalities) involved a car and another motor vehicle, 39 pedestrian (71% of pedestrian deaths) and 12 cyclist (46% of cyclist deaths) deaths involved a car, 9 pedestrian and 6 cyclist deaths involved a truck and 1 cyclist death involved a motorcycle (72–74). A total of 320 of 455 (70%) deaths involved motor vehicle collisions then, so the impact of the motor vehicle on all types of mobility is a major issue. A further issue noted in the report is that 75% of the road deaths are men and that this is only partly explained by distance travelled, with 'a big part is still unexplained' (7).

MacGregor (2002) compares the safety visions of Canada and Sweden. Sweden's Vision Zero ultimately aims to eliminate serious injury and death resulting from accidents and involves goals such as the reduction of fatalities to 250 by 2007. Accidents as well as their human consequences are considered in Vision Zero (135). Canada's Vision 2001 only involved the loosely stated aim that most accidents are avoidable with common-sense solutions such as being hard on drink driving and excessive speeding and enforcing the wearing of seat belts. Canada, with an estimated rate of 9.3 deaths per 100,000 in 1999 (WHO 2004, Table A.4, 192) uses enforcement, fines and harsher treatment of repeat offenders to deal with 'irresponsible driving behaviour'. Sweden's allowable blood alcohol level of 0.02 is lower than Canada's which is 0.08 but Sweden does not just rely on criminal law to deal with the problem. An overall policy on the social institution of drinking as causing drink driving is considered in Vision Zero and has lead to 'countermeasures in drinking policy – reduce drinking – and transportation policy – reduce driving' (135). In Sweden's Vision Zero the road transport system including wider social and cultural practices is considered and the aim is to change the design of the system so that mistakes do not lead to disastrous consequences.

MacGregor argues that many of the pronouncements in Canada's Vision 2001 are more about 'public performance than saving lives' (136). Pronouncements regarding speed for example are not backed up with increased enforcement but focus on individual attitudes and altering individual behaviour relying on the increased safety of motor vehicles. Not only is the lack of safety of SUVs, for example, ignored, motor vehicle safety features generally favour those inside a motor vehicle and MacGregor notes that Canadian policy privileges the rights of motorists over other road users (137). In Sweden on the other hand, an emphasis on equality in Swedish society generally 'encourages traffic-safety strategies that embrace the needs of all road users: the young and the old, rich and poor, motorcyclists, pedestrians, bicyclists and motorists alike' (137). Safety demands are thus set by vulnerable road users and high speeds only permitted on roads where they are not likely to pose a danger to other road users and speeds can be 'technologically controlled'. Nevertheless it is not just road users who are held accountable for deaths but also the transport system itself. This appears to relate mostly to road design and planning although automobile associations are reportedly onside (139). Other areas will need to be considered in the future such as the demands of manufacturers for higher speeds and the promotion of speed and aggression in advertising.

All of the visions considered above have involved different forms of governance. In Singapore governance is more informalised and enlists different sectors of the community in carrying out campaigns. It is also positive and asks something of people that recognises the sociality of the context rather than opposing individuals and authority by merely threatening them with the law. In Sweden, public consultation, focus groups, negotiation, dialogue and debate are used to enlist support and involve the community in their future direction (Forstorp 2006, 110). In Canada the law demands that people obey it while living under the illusion that they are free except for such directives.

The relative acceptability of forms of governance must be related to mobility needs and equality as well as impact on quality of life of current car dependence and regulation. As I argued in the previous chapter, internalised governance is not more insidious because it is internalised, rather it is the forms and inequalities, inclusions and exclusions of governance, whether internal or external that is important. New questions must be asked of statistics as well as current policy relating to how they contribute to the future and what kind of future they are working towards.

A form of reflective self-control that is social and thus able to include others is more valuable than a self-control that is opposed to the social context and thus involves holding down feelings and repressing true desires. Self-control can then be seen as a movement of self within a traffic of others, as working for the benefit and development of the self and freedom rather than opposed to others who are seen as taking away the freedom of the self.

Conclusions

I have flagged a number of important areas in the book that I think need to be the focus of more attention and critique. Apart from the importance of considering the broader social and cultural framework of mobility, particularly driving as a social and cultural and not merely an individual practice, specific issues that have been addressed include gender, responsibilities of manufacturers, the importance of advertising as a framing mechanism, the inappropriate emphasis on racing and rally driving as a model for driving on the roads, and the overemphasis on increasing speed at significant cost to life as well as quality of life. Addressing emotional engagement in the context of sociality has also been highlighted. Where the emphasis is on speed at any cost, experiences of anger and boredom are generated within the everyday functioning of the system and have to do not only with the violence of the car and expectations of instantaneous time but also the affordances of the driver-car. The firepower of the car is tied in with combustion masculinity and more and more of this power is desired to maintain it. Bigger faster roads are not the answer, rather different vehicles and different driver-car assemblages providing alternative models of driving and being for mobile agents.

An analysis of data from the WHO (2004) World report on road traffic injury prevention (Table A.4, 188–195) shows that between 66% and 90% of fatal road deaths are men. Men certainly do more of the driving but they are also more likely to place themselves at risk and to hold values that place higher value on speed over caution and safety or concern for others for example. It is not possible to estimate how many deaths can be attributed to male operators of motor vehicles but at least some of the female deaths on the road are a result of male behaviours. The car is built and promoted to endorse and express those values and when combined with a driver who sees themselves in relation to the vehicle and its power and performance and his ability to demonstrate that, becomes a lethal weapon that supposedly caters to men's desires. Men pride themselves on being skillful in handling cars but are responsible for more of the deaths on the road since they are the ones who drive more. Level of exposure is not an explanation unless death by motor vehicle continues to be considered inevitable. It is inevitable under current structures, expectations and demands.

This vision of mobility engineered by consumer culture is not appropriate for safer, more economical, environmentally and socially friendly mobility. Even 'green' has to be 'mean' according to manufacturers exhibiting at the 2008 Detroit auto show to the relief of thousands of journalists (Dowling 2008). Apparently it is important that carmakers still build performance cars and in fact they 'have no choice' since demand for them remains. Which comes first, the car or the driver? Demand and available styles and types of vehicles are interlinked – one does not exist without the other. It has to change somewhere.

Bibliography

Aberg, L. (1998) Traffic rules and traffic safety, *Safety Science*, 29, 205–215.

Ahmed, Sara (2004) Collective feelings or, the impressions left by others, *Theory, Culture and Society,* 21(2), 25–42.

Aird, Alisdair (1972) *The Automotive Nightmare*, London: Hutchinson.

Allon, Fiona (2004) An ontology of everyday control, In Couldry, N. and McCarthy, A. (eds), *Media/Space: Place, Scale and Culture in a Media Age*, Routledge, London and New York, 253-274.

Ameratunga, Shanthi, Hijar, Martha and Norton, Robyn (2006) Road-traffic injuries: Confronting disparities to address a global-health problem, *The Lancet*, 367, 1533–1540.

Anable, Jillian and Gatersleben, Birgitta (2005) 'All work and no play? The role of instrumental and affective factors in work and leisure journeys by different travel modes', *Transportation Research Part A*, 39, 163–181.

Anderson, Craig L., Winn, Diane G. and Agran, Phyllis F. (1999) Differences in pickup truck and automobile driver–owners, *Accident Analysis and Prevention* 31, 67–76.

Ang, Ien (1996) *Living Room Wars*, London: Routledge.

Appleyard, D.K. (1981) *Liveable Streets*, Berkeley, CA: University of California Press.

Arnett, J.J. (2002) Developmental sources of crash risk in young drivers, *Injury Prevention*, 8(3), 17–23.

Arnett, Jeffery (1992) Socialization and adolescent reckless behavior: A reply to Jessor, *Developmental Review*, 12, 391–409.

ATSB (2003) Potential benefits and costs of speed changes on rural roads, *Road Safety Research Report CR216*, Canberra, Australian Transport Safety Bureau.

Augé, Marc (1995) *Non-places: Introduction to an Anthropology of Supermodernity*, London: Verso.

Barbelet J.M. (2001) *Emotion, Social Theory and Social Structure*, Cambridge: Cambridge University Press.

Barbalet, J.M. (1999) Boredom and social meaning, *British Journal of Sociology*, 59, 4, 631–646.

Barthes, Roland (1973) *Mythologies*, London: Paladin.

Benfield, Jacob A., Szlemko, William J. and Bell, Paul A. (2007) Driver personality and anthropomorphic attributions of vehicle personality related to reported aggressive driving tendencies, *Personality and Individual Differences*, 42, 247–258.

Berg, H.Y. (1994) Lifestyle, Traffic and Young Drivers, *An Interview Study*, VTI–rapport, No. 389A, Sweden.

Bohm, Steffen, Jones, Campbell, Land, Chris and Paterson, Mat (2006) Introduction: Impossibilities of automobility, *Sociological Review*, 54(s1), 3–16.

Bonham, Jennifer (2006) Transport: disciplining the body that travels, *Sociological Review*, 54 (s1), 57–74.

Broomham, Rosemary (2001) *Vital Connections: A History of NSW Roads from 1788*, Sydney: Hale and Iremonger.

Butler, Judith (2004) *Undoing Gender*, New York: Routledge.

Cadogen, John (2004) Rubbery figures, *Wheels Magazine*, February, 28–29.

Carrabine, Eamonn and Longhurst, Brian (2002) Consuming the car: Anticipation, use and meaning in contemporary youth culture, *Sociological Review*, 50 (2), 181–196.

Christensen, P. and O'Brien, M. (eds) (2003) *Children in the City: Home, Neighborhood and Community*, London: Routledge Falmers.

Clarke, David D., Ward, Patrick and Truman, Wendy (2005) Voluntary risk taking and skill deficits in young driver accidents in the UK, *Accident analysis and prevention*, 37, 523–529.

Cockburn, Cynthia and Ormrod, Susan (1993) *Gender and Technology in the Making*, London: Sage.

Cohn, Lawrence D., Macfarlane, Susan, Yanez, Claudia & Imai, Walter K. (1995) Risk-Perception: Differences between adolescents and adults, *Health Psychology*, 14(3), 217–222.

Couldry, Nick (2004) Theorising media as practice, *Social Semiotics*, 14, 2, 115–132.

Couldry, Nick (2000a) *The Place of Media Power: Pilgrims and Witnesses of the Media Age*, London: Routledge.

Couldry, Nick (2000b) *Inside culture: Reimagining the Method of Cultural Studies*, London: Sage.

Crawford, Ian (2007) *Blue Mountains Gazette Review*, March/April, 18.

Csiksentmihalyi, Mihaly (1990) *Flow: The Psychology of Optimal Experience*, New York: Harper and Row.

Dahlen, Eric. R, Martin, Ryan. C., Ragan, Katie, and Kuhlman, Myndi M. (2005) Driving anger, sensation seeking, impulsiveness, and boredom proneness in the prediction of unsafe driving, *Accident Analysis and Prevention*, 37, 341–348.

Dant, Tim (2004) The Driver–car, *Theory, Culture and Society*, 21, 4/5, 61–79.

Davison, Graeme (2004) *Car Wars: How the Car Won our Hearts and Conquered our Cities*, Sydney: Allen and Unwin.

De Certeau, M. (1984) *The Practice of Everyday Life*, Steve Randall (trans) Berkeley, CA: University of California Press.

DeJoy, David (1992) An examination of gender differences in traffic accident risk perception, *Accident Analysis and Prevention*, 24(3), 237–246.

Dowling, Joshua (2008) Red–bloods among the green, *Sydney Morning Herald*, Weekend Edition, January 19–20, Drive, 8–9.

Douglas, Mary (1992) *Risk and Blame: Essays in Cultural Theory*, London: Routledge.

Dula, C S. and Geller, E S. (2003). Risky, aggressive, or emotional driving: Addressing the need for consistent communication in research, *Journal of Safety Research*, 34, 559–566.

Edinsor, Tim (2003) M6-Junction 19–16: Defamiliarising the mundane roadscape, *Space & Culture*, 6(2), 151–168.

Elias, Norbert (1995) Technization and civilization, *Theory, Culture and Society*, 12(3), 7–42.

Ewing, Reid, Schrieber, Richard and Zegeer, Charles (2003) Urban sprawl as a risk factor in motor vehicle occupant and pedestrian fatalities, *American Journal of Public Health*, 93/9, 1541–1545.

Eyerman, Ron and Löfgren, Orvar (1995) Romancing the road: Road movies and images of mobility, *Theory, Culture and Society*, 12, 53–79.

Featherstone, Mike (2005) Automobilities: An introduction, In Mike Featherstone, John Urry and Nigel Thrift (eds) *Automobilities*, London: Sage, 1–24.

Ferguson, S.A. (2003) Other high-risk factors for young drivers – how graduated licensing does, doesn't, or could address them, *Journal of Safety Research*, 34, 71–77.

Ferguson, Susan A., Hardy, Andrew P. & Williams, Allan F. (2003) Content analysis of television advertising for cars and minivans: 1983–1998, *Accident Analysis and Prevention*, 35, 825–831.

Forstorp, Per-Anders (2006) Quantifying automobility: speed, 'Zero Tolerance' and democracy, *Sociological Review*, 54(s1), 93–112.

Fosser, S. and Christensen, P. (1998) *Car Age and the Risk of Accidents*, Report 386/1988 Institute of Transport Economics, Oslo, Norway.

Foucault, Michel (1991) Governmentality, In G. Burchell, C. Gordon, & P. Miller (eds), *The Foucault Effect: Studies in Governmentality*, 87–105, Chicago, IL: University of Chicago Press.

Freund, Peter S. and Martin, George T. (2002) Risky vehicles, risky agents: Mobility and the politics of space, movement and consciousness, In J. Peter Rothe (ed.), *Driving Lessons: Exploring Systems That Make Traffic Safer*, Edmonton: The University of Alberta Press, 105–120.

Freund, Peter S. and Martin, George T. (1993) *The Ecology of the Automobile*, Montreal: Black Rose Books.

Frow, John and Morris, Meaghan (1993) *Australian Cultural Studies: A Reader*, St Leonards: Allen and Unwin.

Gehl Architects (2004) *Places for People*, City of Melbourne Project (available at http://www.gehlarchitects.dk/images/melbourne_2004.pdf).

Graves-Brown, Paul (1997) From highway to superhighway: The sustainability, symbolism and situated practices of car culture, *Social Analysis*, 41(1).

Gregersen, Nils P. (1996) Young driver's over-estimation of their own skill – an experiment on the relation between training strategy and skill, *Accident Analysis and Prevention*, 28(2), 243–250.

Hage, Ghassan (2000) On the ethics of pedestrian crossings or why 'mutual obligation', *Meanjin*, 4, 27–37.

Hagman, Olle (2006) Morning queues and parking problems: On the broken promises of the automobile, *Mobilities*, 1(1), 63–74.

Hakamies-Blomqvist, Liisa (2003) *Aging Europe: The Challenges and Opportunities for Transport Research: Fifth European Transport Safety Council Lecture*, Available at http://www.etsc.be/documents/5th%20European%20Transport%20Safety%20Lecture.pdf (accessed 14.01.2008).

Harré, N., Forest, S. and O'Neill, M. (2005) Self-enhancement, crash-risk optimism and the impact of safety advertisements on young drivers, *British Journal of Psychology*, 96(2), 215–230.

Harré, Niki (2000) Risk evaluation, driving and adolescents: A typology, *Developmental Review*, 20, 206–226.

Harré, Niki, Field, Jeff and Kirkwood, Barry (1996) Gender differences and areas of common concern in the driving behaviors and attitudes of adolescents, *Journal of Safety Research*, 27(3), 163–173.

Horswill, Mark S. and Coster, Martin E. (2002) The effect of vehicle characteristics on drivers' risk-taking behaviour, *Ergonomics*, 45(2), 85–104.

Jain, Sarah S. Lochlann (2004) 'Dangerous instrumentality': The bystander as subject in automobility, *Cultural Anthropology*, 19(1), 61–94.

Jessor, Richard (1987) Risky driving and adolescent problem behaviour: an extension of problem behaviour theory, *Alcohol, Drugs and Driving*, 3, 1–11.

Jessor, Richard (1992) Risk behavior in adolescence: A psychosocial framework for understanding and action, *Developmental Review*, 12, 374–390.

Jonah, Brian A. (1997) Sensation seeking and risky driving: A review and synthesis of the literature, *Accident Analysis and Prevention*, 29(5), 651–665.

Jonasson, Mikael (1999) The ritual of courtesy – creating complex or unequivocal places?, *Transport Policy*, 6, 47–55.

Kellner, Douglas (1995) *Media Culture: Cultural Studies, Identity and Politics Between the Modern and the Postmodern*, London: Routledge.

King, Y. and Parker, D. (2008) Driving violations, aggression and perceived consensus, *Revue européenne de psychologie appliquée*, 58, 43–49.

Laapotti, S., Keskinen, E. and Rajalin, S. (2003) Comparison of young male and female drivers' attitude and self-reported traffic behaviour in Finland in 1978 and 2001, *Journal of Safety Research*, 34, 579–587.

Laapotti, S., Keskinen, E., Hataaka, M. and Katila, A. (2001) Novice drivers' accidents and violations – a failure on higher or lower hierarchical levels of driving behaviour, *Accident Analysis and Prevention*, 33, 759–769.

Langford, Jim, Methorst, Rob and Hakamies-Blomqvist, Liisa (2006) Older drivers do not have a high crash risk – A replication of low mileage bias, *Accident Analysis and Prevention*, 38, 574–578.

Larson, Jonas (2005) Families seen sightseeing: Performativity of tourist photography, *Space & Culture*, 8(4), 416–434.

Latour, Bruno (1993) *We Have Never Been Modern* (translated by Catherine Porter), Cambridge, Massachusetts: Harvard University Press.

Lazar, Michelle (2003) Semiosis, social change and governance: A critical semiotic analysis of a national campaign, *Social Semiotics*, 13, 2, 201–221.

Lonczak, Heather S., Neighbors, Clayton and Donovan, Dennis M. (2007) Predicting risky and angry driving as a function of gender, *Accident Analysis and Prevention*, 39, 536–545.

Lupton, Deborah (1999) Monsters in metal cocoons: 'road rage' and cyborg bodies, *Body and Society*, 5 (1), 57–72.

Lyng, Stephen (1990) Edgework: A social psychological analysis of voluntary risk taking, *American Journal of Sociology*, 95(4), 851–886.

MacGregor, David (2002) Sugar bear in the hot zone, J. Peter Rothe (ed.), *Driving Lessons: Exploring Systems That Make Traffic Safer*, Edmonton: The University of Alberta Press, 125–142.

Markussen, R. (1995) Constructing easiness, In S. Star (ed.), *The Cultures of Computing*, Oxford: *Sociological Review*/Blackwell.

Marsh, Peter and Collett, Peter (1986) *Driving passion*, London: Jonathan Cape.

Maxwell, Simon (2001) Negotiations of car use in everyday life, In Daniel Miller (ed.), *Car Cultures,* Oxford, New York: Berg, 203–222.

McKnight, James A. and McKnight, A. Scott (2003) Young novice drivers: careless or clueless?, *Accident Analysis and Prevention*, 35, 921–925.

McLuhan, Marshall (1964) *Understanding Media, The Extensions of Man*, London: Routledge.

Merriman, Peter (2005) Driving places: Marc Augé, non–places, and the geographies of England's M1 Motorway, In Mike Featherstone, Nigel Thrift and John Urry (eds), *Automobilities*, London: Sage, 145–168.

Michael, Mike (1998) Co(a)gency and the car: Attributing agency in the case of 'road rage', In Brita Brenna, John Law and Ingunn Moser (eds), *Machines, Agency and Desire*, Centre for Technology and Culture, 125–141.

Michael, Mike (2001) The invisible car: The cultural purification of road rage, In Daniel Miller (ed.), *Car Cultures*, Oxford: Berg, 59–80.

Miller, Daniel (2001) *Car Cultures*, Oxford, New York: Berg.

Moller, Mette (2004) An explorative study of the relationship between lifestyle and driving behaviour among young drivers, *Accident Analysis and Prevention*, 36, 1081–1088.

Moore, Clover (2008) Cycling the way to go in this overcrowded city, Opinion, *Sydney Morning Herald*, Friday January 11, 11.

Morphett, Anne and Sofoulis, Zoë (2005) Cars, sex, drugs and media: Comparing modalities of road safety and public health messages, In Lisa Dorn, (ed.), *Driving Behaviour and Training Vol 2: Human Factors in Road and Rail Transport*, Aldershot: Ashgate Publishing, 61– 78.

Morris, Meaghan (1998) *Too Soon, Too Late,* Bloomington, IN: Indiana University Press.

Morse, Margaret (1990) An ontology of everyday distraction: The freeway, the mall and television, In Patricia Mellencamp (ed.), *The Logics of Television*, Bloomington, IN: Indiana University Press, 193–221, 193–194.

Mullan, Elaine (2003) Do you think your local area is a good place for young people to grow up? The effects of traffic and car parking on young people's views, *Health and Place*, 9, 351–360.

Neighbors, C., Vietor, N.A., Knee, C.R., (2002) A motivational model of driving anger and aggression, *Personality and Social Psychology, Bulletin*, 28, 324–335.

Nesbit, Sundé M., Conger, Judith C. and Conger, Anthony J. (2007) A quantitative review of the relationship between anger and aggressive driving, *Aggression and Violent Behavior,* 12, 156–176.

O'Connell, Sean (1998) *The Car in British Society: Class, Gender and Motoring 1896–1939*, Manchester: Manchester University Press.

Ormrod, Susan (1994) Let's nuke the dinner, In Cockburn, C. and Fürst–Dili'c, Ruza (eds), *Bringing Technology Home: Gender and Technology in a Changing Europe*, Buckingham: Open University Press, 42–58.

Packer, Jeremy (2003) Disciplining mobility: Governing and safety, In Jack Z. Bratich, Jeremy Packer and Cameron McCarthy (eds), *Foucault, Cultural Studies and Governmentality*, Albany: State University of New York Press.

Parker, D., Lajunen, T., and Summala, H. (2002) Anger and aggression among drivers in three European countries, *Accident Analysis and Prevention*, 34, 229–235.

Parker, D., Lajunen, T., and Stradling, S.G. (1998) Attitudinal predictors of interpersonally aggressive violations on the road, *Transportation Research Part F*, 1, 11–24.

Parker, Diane; Stradling, Stephen G. and Manstead Anthony S.B. (1996) Modifying beliefs and attitudes to exceeding the speed limit: An intervention study based on the theory of planned behaviour, *Journal of Applied Social Psychology*, 26(1), 1–19.

Parker, Diane; Manstead, Anthony S.B. and Stradling, Stephen G., (1995) Extending the theory of planned behaviour: The role of personal norm, *British Journal of Social Psychology*, 34, 127–137.

Parker, Diane, Reason, James T., Stradling, Stephen G. and Manstead, Anthony S.B. (1995) Driving errors, driving violations and accident involvement, *Ergonomics*, 38(5), 1036–1048.

Parker, Diane; Manstead Anthony S.B.; Stradling, Stephen G.; Reason, James T. and Baxter, James S. (1992) Intention to commit driving violations: An application of the theory of planned behaviour, *Journal of Applied Psychology*, 77(1), 94–101.

Pickett, Charles (1998) Car fetish, In Charles Pickett (ed.), *Cars and Culture: Our Driving Passions*, Sydney: HarperCollins, 23–39.

Pollard, Jack (1974) *Great Motoring Stories of Australia and New Zealand*, Adelaide: Rigby Limited.

Ranney, Thomas A. (1994) Models of driver behaviour: a review of their evolution, *Accident Analysis and Prevention*, 26(6), 733–750.

Reason, J.T., Manstead, A.S.R., Stradling, S.R., Parker, D and Baxter, J.S. (1991) The social and cognitive determinants of aberrant driving behaviour, *Transport and Road Research Laboratory Research Report*, 253. Crowthorne: TRL, 65.

Reckwitz, Andreas (2004) Toward a theory of social practices: a development in culturalist theorizing, *European Journal of Social Theory*, 5(2), 243–263.

Redshaw, Sarah (2007) Articulations of the car: the dominant articulations of racing and rally driving, *Mobilities*, 2(1), 121–141.

Redshaw, Sarah (2006) Driving cultures: Cars, young people and cultural research, *Cultural Studies Review*, 12(2), 74–89.

Redshaw, Sarah (2005) Developing control amongst young drivers, *Youth Studies Australia*, 24, 3, 37-41.

Redshaw, Sarah and Noble, Greg (2006) *Mobility, Gender and Young Drivers: Second report of the Transforming Drivers study of young people and driving*, Sydney: NRMA Motoring and Services, Available at http://www.uws.edu.au/centre_for_cultural_research/ccr/publications#5.

Reith, Gerda (2007) Gambling and the contradictions of consumption: a genealogy of the "pathological" subject, *American Behavioural Scientist*, 51(1), 33–55.

Rose, Nikolas (1999) *Powers of Freedom: Reframing Political Thought*, Cambridge: Cambridge University Press.

Rose, Nikolas (1999) *Governing the Soul: the Shaping of the Private Self*, London: Free Association Books.

Rothe, J. Peter (2002) Red-light cameras: techno-policing at the crossroads, In J. Peter Rothe (ed.), *Driving Lessons: Exploring Systems That Make Traffic Safer*, Edmonton: University of Alberta Press, 291–312.

Rothe, J. Peter (1994) *Beyond Traffic Safety*, New Brunswick: Transaction Publishers, 96–103.

Rundmo, Torbjorn and Iversen, Hilde (2003) Risk perception and driving behaviour among adolescents in two Norwegian countries before and after a traffic safety campaign, *Safety Science,* 42(1), 1–21.

Sarkar, Sheila and Andreas, Marie (2004) Acceptance of and engagement in risky driving behaviors by teenagers, *Adolescence,* 39(156) 687–700.

Scharff, Virginia (1991) *Taking the Wheel: Women and the Coming of the Motor Age,* Albuquerque: University of New Mexico Press.

Schnapp, Jeffery T. (1999) Crash (speed as engine of individuation), *Modernism/Modernity,* 6, 1, pp.1–49.

Sheller, M. and J. Urry (2003) Mobile transformations of 'public' and 'private' life, *Theory, Culture and Society,* 20(3): 107–125.

Sheller, Mimi (2005) Automotive emotions: feeling the car, In Mike Featherstone, John Urry and Nigel Thrift (eds), *Automobilities,* London: Sage, 221–242.

Sheller, Mimi and Urry, John (2000) The city and the car, *International Journal of Urban and Regional Research,* 24(4), 737–57.

Shove, E. (2003) *Comfort, Cleanliness and Convenience: the Social Construction of Normality,* Oxford: Berg.

Silcock, David, Smith, Kim, Knox, Duncan and Beuret, Kristine (2000) *What Limits Speed? Factors That Affect How Fast We Drive,* AA Foundation for Safety Research UK.

Silverstone, R. (1994) *Television and Everyday Life,* London: Routledge.

Simons-Morton, B., Lerner, N. and Singer, J. (2005) The observed effects of teenage passengers on the risky driving behaviour of teenage driver, *Accident Analysis and Prevention,* 37, 973–982.

Sloop, John M. (2005) Riding in cars between men, *Communication and Critical/Cultural Studies,* 2(3), 191–213.

Smith, Alexandra (2008) Empty cycling lanes cost millions: NRMA, *Sydney Morning Herald,* Thursday January 10, 1 and 4.

Sofoulis, Zoe; Noble, Greg and Redshaw, Sarah (2005) *Youth Media and Driving Messages, Transforming Drivers Report 1A,* Sydney: NRMA Motoring and Services, Available at http://www.uws.edu.au/centre_for_cultural_research/ccr/publications#5.

Spinks, Jez, Dowling, Joshua, Hudson, Jaedene and Blackburn, Richard (2008) The road ahead, *Sydney Morning Herald,* Weekend Edition January 5–6, Drive, 1, 6 and 8–9.

Spinks, Jez (2007) Godzilla returns, *Sydney Morning Herald,* Weekend Edition December 8–9, Drive, 8–9.

Steg, Linda (2004) Car use: lust and must, Instrumental, symbolic and affective motives for car use, *Transportation Research Part A,* 39, 147–162.

SIKA (2006) *Sweden Road Traffic Injuries 2006,* Statens institut för kommunikationsanalys, SIKA Statistik Vägtrafikskador 2007:30. http://www.sika-institute.se/Doclib/2007/SikaStatistik/ss_2007_30.pdf.

Thomsen, Thyra Uth (2005) Parent's construction of traffic safety: children's independent mobility at risk?, In Thyra Uth Thomsen, Lisa Drews Nielsen and Henrik Gudmundsson (eds), *Social Perspctives on Mobility,* Aldershot: Ashgate, 11–28.

Thrift, Nigel (2005) Driving in the city, In Mike Featherstone, John Urry and Nigel Thrift (eds), *Automobilties,* London: Sage, 41–59.

Ulleburg, Pål (2004) Social influence from the back seat: factors related to adolescent passengers' willingness to address unsafe drivers, *Transportation Research Part F*, 7, 17–30.

Urry, John (2006) Inhabiting the car, *Sociological Review*, 54(s1), 17–31.

Urry, John (2004) The 'system' of automobility, *Theory, Culture and Society*, 21(4), 25–39.

Urry, John (2005) The 'system' of automobility, In Mike Featherstone, Nigel Thrift, and John Urry, (eds), *Automobilities*, London: Sage.

Urry, John (2000) *Sociology Beyond Societies: Mobilities for the Twenty-First Century*, London: Routledge.

Valverde, M. (1998) *Diseases of the Will: Alcohol and the Dilemmas of Freedom*, Cambridge: Cambridge University Press.

Venkatasawmy, Rama; Simpson, Catherine and Visosevic, Tanja (2001) From sand to bitumen, from bushrangers to 'bogans': mapping the Australian road movie, *Journal of Australian Studies*, December, 75–84.

Vick, Malcolm (2003) Danger on the roads! Masculinity, the car, and safety, *Youth Studies Australia*, 22(1), 32–37.

Virilio, Paul (2005) *Negative Horizon*, London: Continuum.

Virilio, Paul (1986) *Speed and Politics: An Essay on Dromology*, New York: Semiotext(e).

Volti, Rudi (2004) *Cars and Culture The Life Story of a Technology*, Baltimore: The Johns Hopkins University Press.

Walker, Linley (1998) Chivalrous masculinity among juvenile offenders in Western Sydney: a new perspective on young working class men and crime, *Current Issues in Criminal Justice*, 9(3), 279–293.

Walker, Linley (1998) Under the bonnet: car culture, technological dominance and young men of the working class, *Journal of Interdiscipinary Gender Studies*, 3(2), 23–43.

Walker, Linley (1999) Hydraulic sexuality and hegemonic masculinity: working-class Men and car culture, R. White (ed.), *Australian Youth Subcultures: On the Margins and in the Mainstream*, Hobart: Australian Clearinghouse for Youth Studies, 178–187.

Walker, Linley, Butland, D. and Connell, R. W. (2000) Boys on the road: masculinities, car culture, and road safety education, *Journal of Men's Studies,* 8(2),153–169.

Walton, D. and Bathurst, J. (1998) An exploration of the perceptions of the average driver's speed compared to perceived driver safety and driving skill, *Accident Analysis and Prevention*, 30(6), 821–830.

Wenzel, Thomas P. and Ross, Marc (2005) The effects of vehicle model and driver behaviour on risk, *Accident Analysis and Prevention*, 37, 479–494.

WHO (2004) *World Report on Road Traffic Injury Prevention*, Geneva: World Health Organisation.

Williams, Allan F. (2003) Views of U.S. drivers about driving safety, *Journal of Safety Research*, 34, 491–494.

Williams, Raymond (1975) *Television: Technology and Cultural Form*, New York: Schocken.

Williamson, T. (2003) The fluid state: Malaysia's national expressway, *Space and Culture*, 6(2), 110–-131.

Wilson, A. (1992) *Culture of Nature*, Oxford: Blackwell.

Wolf, Winifred (1996) *Car Mania: A Critical History of Transport*, Chicago: Pluto Press.
Wright, Chris and Curtis, Barry (2005) Reshaping the car, *Transport Policy*, 12, 11–22.

Websites Accessed

Aggressive Driving: Where you live matters, http://www.transact.org/report. asp?id=58
Department for Transport UK (2005) Home Zones: Challenging the future of our streets, http://www.dft.gov.uk/stellent/groups/dft_susttravel/documents/ divisionhomepage/610453.hcsp
Department for Transport UK (2000) New Directions in Speed Management: a review of policy, http://www.dft.gov.uk/stellent/groups/dft_rdsafety/documents/ page/dft_rdsafety_504682.hcsp
Kubilins, Margaret Designing Functional Streets that contribute to our Quality of Life, http://www.urbanstreet.info/1st_symp_proceedings/Ec019_b3.pdf
Traffic's Human Toll http://www.transalt.org/campaigns/reclaiming/reports.html

Index